BORN TO FARM

By

David Black

Grosvenor House
Publishing Limited

The right of David Black to be identified as the author of this
work has been asserted in accordance with Section 78
of the Copyright, Designs and Patents Act 1988

Book cover © David Black
Front cover image © Richard Flagg

This book is published by
Grosvenor House Publishing Ltd
Link House
140 The Broadway, Tolworth, Surrey, KT6 7HT.
www.grosvenorhousepublishing.co.uk

A CIP record for this book
is available from the British Library

ISBN 978-1-83975-991-8 (Hbk.)
ISBN 978-1-83975-931-4 (Pbk.)

CONTENTS

FOREWORD

'Born to Farm' is such an apt title for this fascinating book which follows the fortunes of one Suffolk farmer and the farm which has been his life for almost ninety years.

From 1898, when Grandfather Black brought the family down from Scotland to farm in Suffolk, to the present day, they have always managed to combine the most up-to-date farming methods with the ability to look ahead, plan carefully and experiment. They have tried almost everything from wheat to cut flowers, from pigs to rhubarb.

I found it particularly absorbing not just because my wife and I have been friends of the Black family for many years, but also because I have had a lifelong interest in farming, its history, and its challenges.

It covers the time I have lived through and reflects on life in the county where I was an MP for 27 years and still live. Most importantly it shows great insight into how farming has arrived at its present, pivotal position in the great environmental debate.

I enjoyed the anecdotes and the occasional literary references. Kipling's 'satiable curiosity' I thought very relevant.

Throughout the book you can feel the restless energy that keeps the author on a continual quest for improvement and spurs him on to travel widely to learn and keep ahead of the game.

Anyone with an interest in the countryside, farming in Suffolk and particularly how they have interacted over the past century, will thoroughly enjoy this book which makes an important contribution to the history of rural life in the county.

Lord Framlingham
Suffolk
November 2021

AUTHOR'S NOTE

I have endeavoured, at the behest of various members of the family, to put on paper some of my remembrances from the past. It has taken four or five years, in fits and starts, in the writing and inevitably there are gaps as I never kept a comprehensive diary. I have recalled what I can with the help of farm records, photographs, correspondence, news articles and conversations. That it might have any coherence at all is due to my daughter, Alison, who has been driving hard behind the scenes, and has spent an enormous amount of time in doing so. My wife, Mary, has been asking for a long time when or even if it might be finished. We have finally got there, and I dedicate this book to her.

I hope 'Born to Farm' sheds a little light upon the enormous changes that have taken place in farming over the past 90 years as seen through the eyes of one of the participants.

INTRODUCTION

Farming is in my genes. In 1898 my grandfather, David Black, began farming the 194-acre Red House Farm in Bacton, Suffolk as a tenant farmer. Both he and my father saw huge changes and developments in farming during their lifetimes. The pace has not slowed since I joined the business aged 17 in 1949. Red House Farm is still the hub of the business, but today, David Black & Son Ltd and its subsidiary, Bacton Pigs Ltd, farms 2000 owned acres and a further 1000 acres under contract, with a property division comprising 50 residential properties and several commercial units for let.

Farming is a fascinating way of life. A farmer is rooted to the spot yet needs to keep abreast of changes in the rest of the country and in the world, not just in terms of market opportunities and technological advances, but regarding political and environmental issues too. I have been fortunate that, for me, it has been as much my hobby as my work.

CHAPTER 1

EARLY MEMORIES

Is it possible to be sure of one's earliest memory? Was mine looking out of the little kitchen window to see Ernie Jackson's steam engine with drum and pitcher behind clunking up to thresh a stack of corn? Or was it being bathed in the old tin tub in the outhouse which served as an extension to the kitchen? Or perhaps it was looking out to see the horse-drawn trailers going to and from the farmyard. I have been told that I called the trailers 'gee-gee carts with pom-pom wheelies', for by then the iron-shod wooden wheels of the old standard road wagons were giving way to pneumatic tyres.

It snowed on 3 April 1932, the day I was born at Somerville[1], a house built for Mother and Father when they married in 1931. It was at the end of the drive leading up to Red House Farm in Bacton, not a very big house, as farming times were hard and money was short. My father, Tom, an only child, had been born in 1900 at Red House, an old, rambling farmhouse, the home of my paternal grandparents. They, David and Kate Black, had left Scotland in 1898 and begun their married life, aged 24 and 22 respectively, renting the 194-acre Red House Farm in mid-Suffolk. They had arrived by train at Finningham station three miles away, bringing with them, at a cost of £17 10s for carriage, a herd of 20 Ayrshire cows, implements and furniture. Over the years, by scrimping, saving and sheer hard work they were able to buy the farm from the landlord, Mr. Fitch, in 1911, and then gradually take on more land.

---

[1] Somerville was Grandma Red House's maiden name. The house name was for a time spelt Summerville.

But to return to my memories, the kitchen at Somerville was a tiny room with an outside door opening to the elements. In it was a triplex stove, a sink, a table and two chairs, and a bench for me and my brother Robert, who was two years my junior, to sit on. This left hardly enough room to turn around. There was, however, a connecting, covered walkway from the kitchen to the outhouse, where Mother would cook on a three-burner paraffin Valor stove, and in an oven which stood on top of one of the burners. On a high shelf above the stove among other paraphernalia Mother stored long bars of soap. The water was heated in the copper, a built-in copper cauldron with a lid to keep the steam in and a wood fire beneath. Water had to be tipped in by bucket and ladled out for use. This was rainwater, collected in tanks from the roof and often had 'wrigglies' in it. Our drinking water came from the hand pump at Red House, one or two buckets of which were brought by Grandma's bacc'us[2] boy every day. Robert and I helped when we were old enough. The WC was in the outhouse, too – a new flush loo. We used cut-up newspaper, hanging from a string on a hook on the back of the door. My other grandmother, Granny Gipping, still had a three-holer outside. It made for a chilly visit in the winter.

Mother was born Jane Charlotte Edwards to Philip and Rosetta Edwards who farmed Rookery Farm, Gipping. As a child, she fed the fowls and collected their eggs, kept rabbits for pin money and could skin a rabbit in a trice. Although brought up to the farming way of life, she went into teaching and had a love of dancing which is how she came to meet Father at a tennis club dance at the TA Drill Hall in Stowmarket in 1930.

Father had left school just before his 16th birthday in 1916 to work on the farm, as labour was in short supply with so many men having been called up by the Army. His

---

[2] Bacc'us boy: a lad who helped 'at the back of the house' doing odd jobs such as collecting eggs, chopping wood, cleaning boots and shoes, sweeping snow off the paths.

father, my grandfather, must have made some impact locally because during the Great War he was called on to give advice on how to produce more from the land, so he was often out visiting other farms, leaving Father in charge. Farmers were being exhorted to grow more food because the German submarines had begun to sink our supply ships causing food production in the UK to become a national concern.

As I grew up, I remember spending a good deal of time at Red House. It was the hub of the farm and housed the office. It was where the salesmen came and where visitors and family were entertained. On a Friday night, the farm men would queue up at the half-door into the kitchen, bringing their time sheets so they could be paid. Grandma would ask what they had been up to that week, enquiring about their work and their families, and pay them their wages.

In my father's arms at Bacton
Flower Show in 1934

With my mother, 1934

Me aged 2 with Ruby Whatling at Red House

Robert, Mother and me, 1935

The half-door at Red House opened onto a large brick-floored kitchen which contained a four-door aga, a copper, a large fireplace and a stone kitchen sink with a hand pump over it for drinking water. The washing was done in the kitchen along with the cooking. When Grandma was baking, she would leave the buns and tarts on the copper to cool. The temptation was usually too much, and Robert and I would often pick up a freshly made jam tart on the way through the kitchen to the office. Sometimes, we would scoop out the jam and then ask for the tart to be refilled. Once a year, Nurse Woodward would bring her honey extractor to spin out the honey from the honeycombs taken from the hives in the back orchard. During World War II it was the place where harvest rations were divided and distributed to the farm workers. Leading off the kitchen was the dairy, used until 1928 when the last of the cows was sold. It had wooden louvres rather than glass in the big window at the back. A flush WC had been installed in one corner of the old dairy to replace the privy which remained, ramshackle and ivy clad, in the back garden.

A long, dark passage gave access to the other rooms. The first room on the left was known as the breakfast room, the most-used room in the house. A fire was always burning in the grate and there were comfortable easy chairs and a table, which not only served for meals but provided space for doing puzzles, playing draughts, building models and all the countless things that we boys and the adults did. On the sideboard was a barograph into which Grandad would insert a fresh chart each week. The highs and lows would provide a source of comment and discussion. Immediately opposite the breakfast room was the staircase.

At the end of the passage, a door led to the dining room, which was used for entertaining. The passage turned to the right leading to the pantry. From there, two steps up from the passage, one entered the office, a room of less than 10ft sq with a window looking out across the moat to the stable and stackyard beyond. Under the window was a desk, and books and ledgers filled the shelves which covered the walls. An Underwood typewriter sat on the office desk next to an old-fashioned upright telephone and its separate earpiece. Originally the farm telephone number was Bacton 3 then 243, and later 781243. If either Grandad or Father were needed, the farm secretary, Maud Wilson, who lived in the house, had to go outside and blow a whistle to attract their attention, which meant of course that callers could be hanging on for quite some time before they had an answer. Those were the days when telephone exchanges were manned and the operator was able to tell you if the person you hoped to call was in or out and maybe even what they were up to. Somerville was on an extension line and calls had to be put through from Red House, a source of annoyance, particularly to Mother, who felt her conversations were often being listened to by her mother-in-law.

Before the advent of the telephone, messaging was done either by letter or telegram. Letters were delivered to an office in the old apple shed before 7 o'clock each morning and there was a second delivery in the afternoon. Telegrams were delivered free of charge within three miles of the Finningham Telegraph Office. Red House was just outside that limit, however, so to save the delivery cost telegrams were taken to Henry Farrow's at Earls Green. One of his family would bring them to Red House. Henry worked on the farm and he provided a steady stream of daughters who worked as maids at Red House. After the Farrow girls married and moved on, Ruby Whatling became the maid, and Robert and I were often entrusted to her tender care. Ruby's father, Freddie, was head horseman and when he retired, his son, Freddie, Ruby's brother, took over the post. When Ruby married, her sister, Sylvia, became maid. We were fortunate in having several loyal families who worked with us over the generations.

Robert and I often spent the night at Red House, initially in the little room on the other side of Grandma and Grandad's bedroom. Maud would sometimes read to us before we went to sleep. Wordsworth's *Daffodils* made a lasting impression on me, maybe because the fields of daffodils which we grew on the farm allowed me to visualise the flowers 'fluttering and dancing in the breeze'. I also loved Hans Christian Andersen's fairy tales especially *The Sandman*, which was about little Hjalmar who had the most wonderful dreams. I wondered if I could have them too. If I remember correctly, the sandman himself came in the evening and threw sand in the eyes of the little ones to put them to sleep; the good children would have wonderful dreams and the less good a dreamless night. Rather like *The Arabian Nights,* which also fascinated me, Hjalmar recounted a succession of his dreams. I well remember summer nights when it was still light and, having been put to bed, looking out of the window and seeing the horses being watered in the moat before they were put out in the Avenue to graze for the night.

One of the upstairs rooms at Red House, known as the apple room, though no longer used as such, was filled with books which Grandad had amassed over the years, and I spent many happy hours browsing the shelves. There were farming books with diagrams of animals, with flaps that could be peeled back layer after layer, revealing bones, muscles, veins and eventually organs; there were history books, geography books, adventure stories. There were also translations of books written in German, French and Russian. He had piles of practical magazines, mostly American, which included articles on how to make radios, tips on using a lathe or on making a wire recorder. Where they all came from, I never knew, but they made fascinating reading and gave me an everlasting love of books.

Like Maud, Aunt Janet was a semi-permanent resident at Red House. She was Grandma's unmarried sister and would spend half the year at Red House and the other half with another sister, Maggie, who had also married a farmer and moved south of the border. They farmed at Layer Hall, Layer-de-la-Haye, near Colchester[3]. Aunt Janet was always knitting, having had instilled in her from an early age the saying that *the devil finds work for idle hands.* One evening, when an air raid was expected, Grandad called her to go into the cellar. 'Och', she said, 'I canna come; I've dropped a stitch.' Fortunately, the nearest bomb to be dropped near Red House during World War II was over a quarter of a mile away. Uncle George, Grandad's brother, and his family who lived and farmed at Potters Farm, Mendlesham, were not so lucky. They heard a German plane fly over before it dropped a bomb which landed within yards of the house, blowing out most of their windows. Thankfully, no one was hurt but it left a big hole in the asparagus field.

I particularly recall Halloween parties at Red House in my early teens. Grandma would invite us, sometimes with another farming family such as the Sheldons from Wyverstone, to join in the fun. We would duck for apples bobbing in water, plant our faces in flour in search of hidden

---

[3] Part of Layer Hall Farm was taken for a reservoir in the 1930s to provide drinking water for the ever-expanding population in north Essex and south Suffolk.

sweets and then try to bite a scone covered in black treacle hanging on a string with our hands behind our backs. This all took place in the kitchen, prior to a special Hallowe'en supper.

Grandad was chairman of the committee that ran the annual Bacton Flower Show, the highlight of the year in the village. It was held on the Saturday preceding August bank holiday Monday, which used to be the first Monday in that month. The show was instituted sometime at the beginning of the century and took place in one of our meadows at the top of the Avenue, which unsurprisingly became known as the Show Meadow. There were marquees and a fair as well as fiercely contested competitions for all classes of vegetables, flowers and cookery. Grandad always entered the gladioli classes and invariably won, as we were growing about 20 acres of these flowers and he could select the best.

The afternoon of the show was given over to sports for all age groups. A popular event was musical cars, rather like musical chairs. The cars all had running boards, and on each car, a girl would stand on the running board while her boyfriend drove the car round in a great circle. Grandad would be on a lorry in the centre of the ring waving a flag. When the flag came down the girls had to jump down from the running boards and rush to stand on one of an ever-diminishing number of sacks beside the lorry. The girl who stood on the last sack was the winner, together with her beau who had been driving the car. I wonder what health and safety would have to say about that today? These events, with richly rewarding prize money, culminated in a pole vault and a mile race. In the evening, the fun fair came into its own; there were steam horses, swing boats, coconut shies and shooting galleries, and the 'peep-peep' of the steam engine driving the roundabout would sound until midnight. The show did not fit in very conveniently with harvest. One year we started carting and threshing some turnip seed at four in the morning so that we could be through by midday for everyone to be free to attend.

During World War II, the show was discontinued but it started again in 1946. That first year after the War there must have been over 2,000 visitors and most of them came on bicycles, hungry for entertainment in a form which had been denied over the War years. I was always amazed at how ordinary the meadow looked on the Sunday morning compared with how vibrant and boisterous it had been just six hours previously; the funfair people would have already packed up everything and moved on to Eye where they provided entertainment for the bank holiday Monday. One year, however, it was particularly wet, and they were still there on the Sunday morning. They were grateful that we could provide a crawler tractor and driver to pull them off out of the mud so they could be on their way.

Mains electricity came to Bacton just after the War in 1946. Until then we used Tilly lamps, hurricane lamps and candles for lighting, to supplement what electricity was produced from a DC generating set which charged up 55 large glass accumulator batteries. Father had bought these from the old workhouse at Dickleburgh when it was sold in 1930. In fact, we had two generators, both driven by diesel engines: the big one was used for grinding corn and mixing meal and at the same time charged the batteries, while the other was a standby for when the batteries needed topping up. The system provided some light and power around the farm buildings. It also provided lights for Red House and Somerville, and I can remember when the lights began to dim, Father would go out to start the auxiliary engine.

Although Grandad had tried out an early motor plough in 1916 and had bought his first tractor in 1918, a 30hp Wallis Cub Junior, working horses were still very much in use on the farm in the 1930s and 1940s, as they were throughout the rest of the country. Much of the ploughing was done with the tractor but the horses did the 'top work', pulling the harrow, the roll, the drill and hoeing the beet. This latter we still called 'horse hoeing' even when we had

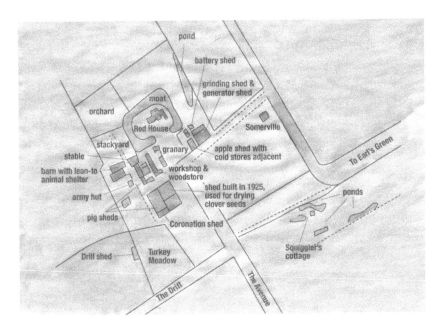

Plan of Red House Farm buildings as in 1936

moved on to using a tractor. The horses had to be shod and every village had its blacksmith. There were two in Bacton. As small boys on the way home from the village school, Robert and I would often call in at Dick Webb's forge beside the pond by the village shop to see a horse being shod or ask if we could have a turn on his bellows keeping the fire going while he heated a horseshoe or some other piece of metal.

Before mains electricity was laid on, we had no welder on the farm so any implement that needed welding would be taken to Dick Webb; he was always happy to have a chat. Dick had learnt the trade of blacksmithing and horse-shoeing while in the Army serving in India. Sometime between the wars he left the Army, came to Bacton, married a local girl and set up his business. The harrow was much needed in the days before chemical weedkillers made their appearance and Dick would sharpen our harrow teeth, heating them in the forge and hammering them out. In the same way he straightened out implements which had bent or twisted and reckoned there was a special place in hell for cold iron blacksmiths. Outside the forge he had a large steel plate and it was on this that he would re-tread wooden wheels by putting a steel rim around their circumference. I found this fascinating to watch. He would make a new band slightly smaller than the diameter of the wooden wheel, then heat it until it glowed red hot. Once expanded, he would place it around the rim of the wooden wheel and then douse it with water from the pond. This made it shrink and tighten onto the wheel, a practical lesson in showing us how metal expanded when heated.

I had some rather embarrassing experiences with the working horses when I was a child, partly because Grandad and Father expected me to be able to cope with any animal even if it was 10 times my size. On one such occasion I had been sent to the stable to saddle up Captain. Being young and not very tall I climbed into the manger so that I could reach to put on his collar, but just as I was about to put it over his head, he gave a snort, tossed his head and I fell back into the manger with the collar over my head instead. I must have been 10 or 11 when I took Ruby, another of our horses, to be shod, on my own for the first time. Fred Whatling set me off from the farm riding bare-backed with only a bridle. When the shoeing was successfully completed, Dick helped me remount and I started for home. Shortly after I had left the forge, a car overtook us and frightened Ruby. She got on the trot and I was unable to slow her down. It was as much as I could do to hold on and, by the time we reached the stable door at Red House, I had a very sore bottom. The next time, I walked her home.

When Grandad had come to Suffolk in 1898, wheat and oats were cut with a reaper and then tied in sheaves by hand. Barley was cut with a scythe and carted loose; it was no wonder the workers wore a neckerchief to keep out the barley havels[4] when they were loading the horse-drawn wagons. A year after taking the farm, Grandad bought the first binder[5] in the area. It was hard work for the horses which pulled it but a great improvement from the old way of harvesting, although most of the locals said it would not last. Change did not come readily to our rural communities. The binder cut the corn and packed it into bundles or sheaves. These were then stood up by hand into shocks or stooks, before being carted to the stackyard. It was said that oats should spend three Sundays in the shock, but seldom did, though wheat could be carted after only one week. This 'waiting' was to allow all the grain to ripen evenly as well as to give time for the weeds to wilt and not cause heating in the stack. Threshing was done in the winter months as and when grain or straw was needed.

Even by the 1950s most of the grain was still cut by binder and put into stacks in the stackyard. Each stack would contain 10 or 12 tons of grain, which would be the produce of five to 10 acres depending on the yield. These stacks would be threshed out later in the year. It would take all day to thresh a stack of corn and it required at least eight men in the gang: two on the stack, one on the drum cutting the bines, two on the straw stack, one bagging chaff, one weighing and bagging the corn, and then the steam engine man, in our case usually Ernie Jackson. His job was to get steam up before the rest of the gang came to work, then during the day to make sure there was enough water and coal to keep the engine running. He had to set up the engine, the drum and the pitcher in alignment so that the driving belts would operate efficiently. I can hear him now calling for three strong lads to shake the pitcher over so that the belt would stay on. It was not surprising that rats and mice and sparrows would find the stackyard a great source of food

---

[4] Havel: the prickly 'awn' or bristle on the grain.
[5] Binder: used for cutting corn and binding it into sheaves This one was a Massey Harris, 5ft machine, which cost £35 in 1899.

Cutting corn with binder,
harvest on Red House Farm, 1940

Shocking wheat, harvest 1940

Loading sheaves, 1940

Threshing at Red House

during the winter, though they in turn would be the target for owls and the farm cats as well as a little boy prowling around with a .22 air gun. When a little older, I was able to help at harvest time. I shall never forget Fred Whatling telling me to 'stamp on 'em boy' if the sheaves of corn arrived at the top of the stack faster than the two of us could place them.

Grandad told me that when he first came to Bacton the fields were so short of phosphate that he could write his name on the fields with superphosphate. What he meant was that, where the superphosphate was applied to the soil, the crops would grow that much better. When he arrived from Scotland, he found that only a little muck and no fertiliser was being used on Suffolk's heavy clay. However, by the time he had drained the land and added nitrates in the form of nitrate of soda and superphosphate he won prizes for producing the best mangolds and sugar beet in the area. Grandad was always experimenting. When spreading fertiliser, he would leave an area of the field without fertiliser, an area with just a little and another with twice as much. This way he was able to judge better the appropriate amount to use in the future. The fertiliser was spread by hand from a sowing sheet. The usual Suffolk practice at that time was to scatter fertiliser or seed with one hand. Grandad, however, had been brought up using both hands and his technique apparently was quite a talking point in the village.

Grandad had many interests apart from practical farming. In 1919, together with three other prominent Suffolk farmers, he founded the Suffolk branch of the National Farmers Union (NFU). For the first two years he was both chairman and secretary, after which a full-time secretary was appointed. He carried on as chairman for the next 10 years and continued to be active at national level, becoming at one stage chairman of both the seeds and pigs committees.

In 1922 Grandad was approached by the manager of Eastern Counties Farmers who said they had a Scottish firm looking for farmers to grow grass seeds for them. That conversation began a long and profitable relationship with McGill & Smith, seedsmen from Ayr. John Watson, one of the directors, was a frequent visitor, coming to see how the cocksfoot and the fescues[6] were growing on the farm and often staying overnight at Red House. He became a family friend and he and Grandad would spend hours searching the hedgerows and verges looking for superior strains of grasses and clovers. They found one cocksfoot plant which Grandad multiplied; McGill & Smith sold it commercially under the label 'Norton' for it was on our Norton Farm that it was found. Sometime previously Grandad had found a particularly large ear of wheat which stood out among others. He took it home and multiplied it. When he had enough, he sold the grain under the name of Bacton Masterpiece. The knowledge and exploitation of genetics have come on apace but back then improvements usually came about by someone spotting a naturally occurring sport.

Another frequent visitor to Red House was Colin Brett, a director of the pig processing side of Wall's[7] and a member of the NFU pigs committee. By that time, we were keeping pigs, and with Grandad's active interest in the local co-operative, St Edmundsbury bacon factory at Elmswell, they had some lively discussions. Wall's was pursuing a different strategy in pig production and processing from that of the factory at Elmswell. To a youngster growing up it was fascinating to hear the discussion and arguments as to what they considered the appropriate way forward for the British pig industry. Something must have rubbed off, because years later I was to become much involved in the pig side of the business.

James Turner, later Lord Netherthorpe, President of the NFU, and later still chairman of Fisons[8], would also call in at Red House to discuss agricultural politics. Grandad

---

[6] Cocksfoot and fescue: species of grass.
[7] Wall's: a food-producing company, now part of Unilever.
[8] Fisons: a company specialising in fertilisers and horticultural chemicals, with headquarters in Ipswich until taken over in 1995 by Rhone-Poulenc.

frequently wrote to the agricultural press, and Malcolm Messer, the editor of *Farmers Weekly*, and Alan Ramsey from *The Farmer and Stockbreeder* were others often at Bacton. Grandad was a regular visitor to the National Institute of Agricultural Botany at Cambridge, taking a keen interest in its development work on cereals, clovers and grasses. He was also close to the work of the National Agricultural Advisory Service, the precursor to ADAS, getting involved in field trials and always learning something new.

As a result of these contacts, Grandma entertained a steady stream of interesting people. She was renowned for her high teas and we would often be invited to eat with notable guests in the dining room at Red House. Cold meats and salads and 'tattie scones'[9] featured as did all manner of homemade buns, tarts and pancakes[10]. Early on Robert and I would crawl about beneath the table undoing shoelaces or perhaps tying them up. Fortunately, we grew out of that, otherwise we would have been in trouble, as Grandad could be a bit sharp if we did not behave. He was a master of the put down, correcting us when he thought it necessary, often using phrases such as, *'Never show fools or children half-finished work', 'Better to be thought a fool than open your mouth and remove all doubt', 'Children should be seen and not heard.'* Not surprisingly we knew our place, but he encouraged us to listen, and we were able to learn and benefit from the discussions we heard.

His membership of various committees and their meetings took Grandad to London and around the country. On his return he liked to sit and chat with Father in the breakfast room to catch up with what had been happening during the day on the farm, and to relate what had been going on at his various meetings. By the time I was working on the farm I often sat in on their discussions, curious to hear about the wider agricultural world.

There were markets in all the local towns when I was growing up and market day was usually the day the week's

---

[9] Tattie scones: scones made from potato.
[10] Pancakes: drop scones.

shopping was done and local gossip exchanged. Market day was a Tuesday in Ipswich, Wednesday in Bury, Thursday in Stowmarket and Saturday in Norwich. Initially Father sold his pigs through one of the two markets at Stowmarket, either Knights or Woodwards, though neither exists today. Very little livestock is now sold through such markets, and it must be 40 or 50 years since cattle and pigs were driven through the streets of Stowmarket on their way to and from the railway station, a commonplace sight as I was growing up. The Corn Exchanges too have gone but I can just remember going with Father to the one in Bury St Edmunds, taking a sample of what he hoped would be malting barley which he offered round to the merchants to see who would give us the best offer.

Father was always very good to us boys. We used to spend the holidays from school running around on the farm whenever we felt like it, helping, or so we thought. He taught us to climb the chestnut trees and the pines just outside Somerville. Left to play on our own, I well remember running around on the roofs of the buildings, climbing straw stacks and building dens out of the wooden flower boxes. I look back and think how lucky we were not to have hurt ourselves more than getting an odd graze or two.

Our first combine harvester, an International 62, had been bought in 1942, though we needed a permit to obtain it. It worked so well that a second was ordered. It arrived the following year at Thurlow's yard at Stowmarket in packing cases, having been shipped from the USA. It was the school holidays; I had just turned 11 and I was mightily proud to be allowed to go with Charlie Hales and Kenny Betts from the farm to work with two of Thurlow's men to help put it together. It took just under a week to assemble; my contribution was to sort out the nuts and bolts. Even though we then had two combines on the farm, we were not able to dispense with the binder until we bought larger combines some five or six years later.

Combines were not the only game changers, for I also remember going from school to the first national sugar beet harvester demonstration at Great Livermere. I was amazed

Combining with International
combine harvester, 1943

at the wonderful contraptions being used, though many of
them were from the US designed for light land so would have
become completely bogged down when our heavier land
became wet. Remarkably, 70 years on and after many
changes and developments, we now have a sugar beet
machine capable of doing what would have needed 50 or 60
men then.

I was always encouraged in what Rudyard Kipling called 'a
satiable curiosity'[11] and was not afraid to ask questions,
reassured, perhaps, by his '*I keep six honest serving men, (They
taught me all I knew), Their names are What and Why and
When, and How and Where and Who*'. So when it was time for
me to go to school, I was full of excitement and expectation,
keen to find answers.

---

[11] *A satiable curiosity*: from Rudyard Kipling's *Just So Stories: The
Elephant's Child* (1902).

My paternal Grandparents'
Golden Wedding Anniversary, 1948
Standing: L-R George Black, unknown,
Grandad, Grandma, Uncle Fred Ford, Father, Uncle George Church
Seated: L-R Margaret Faulds, Olive Black
(wife of George), Aunt Janet Somerville (sister to Grandma), Aunt Nan
Ford (Nee Agnes Black), brother Thomas (age 2), Mother, Auntie Mary
Church
On ground at front: L-R: Robert, David (me)

At Dunnet Head, Caithness, with Mother and Father in 1953.
The car is a Riley. When Mary and I started courting,
she called it the Rattley Riley

Grandfather and Father with Alan Ramsay,
editor of the Farmer and Stockbreeder,
at Red House, 1952

Somerville (photograph taken in 1961.
The extension on the right-hand side
was built in 1950)

CHAPTER 2

SCHOOL DAYS

I started in the infants' class at Bacton School aged four and a half in 1936. The village school then had just four classes, teaching children all the way up to the age of 14. When I first went to school, I was able to ride a bicycle but had difficulty getting on and off. Mother would help me on to my bicycle at home and I would cycle to school. Once there, I would have to circle around until somebody helped me off before I mastered the art of dismounting by myself after a month or so. In the winter, when there was snow on the ground, I would walk. If I was lucky, Father would be going to the Wickham or Mendlesham farm and he would take me in the car.

The headmaster of the school, George Church, was my uncle by marriage to Grandad's half-sister, Mary. They had first met when Auntie Mary came to live at Red House after her parents, Grandad's father and stepmother, emigrated to Australia with their other children at the beginning of World War I and she chose not to go with them. George Church had begun his teaching career at Wenhaston[1], Suffolk, and was appointed headmaster of Bacton School on the death of his father, the previous head. The schoolhouse went with the job. When brother Robert joined me at the school, Mother arranged that Auntie Mary would provide us with lunch in their home. Uncle George was a great stimulant, full of ideas, and encouraged us to play all manner of games. He

---

[1] Wenhaston: a small village in north-east Suffolk, approximately seven miles from Southwold.

kept a pair of stilts in an outside shed and showed us how to walk on them. More cerebrally, one game I particularly remember playing some lunchtimes was having to take turns in working through the alphabet naming sometimes animals, sometimes birds, another time flowers. Uncle George was also a keen draughts player, as indeed were my parents. We were taught the rudiments of the game early on but, try as I might, even as I grew older the best I could do was to force a draw, never managing to beat Uncle George.

With Mother, we would often cycle to Grandma Edwards's at Gipping, some six miles away. Sometimes the roadmen would be tarring the road and covering the hot tar with sand. It was a messy job and we had to be careful not to get the tar on the bicycle wheels. On one occasion I was permitted to cycle to Grandma's on my own, probably aged five. By this stage, I could get off my bike but could not yet get back on unaided. On the way, I needed to relieve myself so got off, but then, unable to remount, I had to push the bicycle. Fortunately, I had not gone far when I found the local roadman who helped me up and I was on my way again.

I suppose we must have spent most of our early years at school learning the three Rs but we also had a great deal of encouragement from home. I remember reading out aloud in class and having to learn our times tables which we did by constant repetition. Much of our learning was by rote, of poems and pieces from the Bible such as, '*I am the true vine and my father is the husbandman*', '*The Lord is my shepherd...*', '*though I speak with the tongues of men and of angels and have not charity...*'. With little prompting I can recite them today.

There was no drinking water at the school so every day two or three boys were sent with a water cart to the village pump beside the railway bridge. I suspect that they spent longer than was strictly necessary, but we then had water to supplement our one-third of a pint of milk which was issued to us at playtime. We did exercises in the playground and, on

Evacuees newly arrived to stay at Red House, 1939

Empire Day[2], we would march round the flagpole and salute the Union Jack flying from it.

In September 1939, the evacuees, about 50 or so children from London, arrived with their schoolteachers and were billeted in the village.

The five girls who came to Red House were from two families who lived in the East End of London. Their schoolmistress was billeted with us at Somerville. I presume they did their schooling in the village hall across the road from the school, for they took their lessons apart from us. The winters of 1939/40 and 1940/41 were particularly cold. We used to walk to school with the girls through the snow. Playing together we built snowmen and a snow fort in the garden at Red House, which lasted long after the rest of the snow had melted. They could only have stayed for 18 months

---

[2] Empire Day: held on 24 May each year, Queen Victoria's birthday, to celebrate the British Empire. I believe they also celebrated Empire Day in Australia and Canada. However, in 1958 it was rebranded as Commonwealth Day.

or less for the anticipated bombing of London had not yet started in earnest and people were calling it a Phoney War. It didn't seem as though the countryside really suited them but maybe I was wrong, for many years later we received a letter from one of the younger girls saying how much she had enjoyed her time at Red House Farm. As part of the nationwide campaign during World War II to encourage people to grow their own food we too joined the Dig for Victory call. It was in my last year at the village school when I was eight years old that we prepared a vegetable garden opposite the school, and I remember digging my patch and planting blackcurrant bushes and potatoes.

The brighter children at Bacton usually took an exam to go to Stowmarket Secondary School. Others would stay on at Bacton School until they were able to leave at 14 to find a job, usually on one of the local farms. Various options were explored by my parents regarding our future education. One day Father was speaking about it with Frank of Frank J Nunn and Sons, our local machinery dealers. He recommended Culford School, near Bury St Edmunds, where he had sent his three sons; he was very pleased with the schooling they had received. Soon after, Father took me and Robert to Culford where we met the headmaster, Dr John Skinner, and were taken round the school. I was given a short test and presumably he was happy with the answers I gave to his mental arithmetic and spelling questions, for Father signed the forms there and then and I started as a boarder in the summer term of 1941.

The termly reports from Culford School gave us grades in each subject and recorded our weight, height and chest expansion. Looking back through those my parents kept, I note that at the end of that first term I weighed 4st 8¾lb, stood 4ft 3¾in tall and my chest expansion was 2 inches. By the time I left school in July 1949 shortly after my 17th birthday I weighed 10st 11lb, stood 5ft 9¾in tall, and my chest expansion had doubled to a healthy 4 inches. Do schools report such information about pupils nowadays?

Not surprisingly, I had my bouts of homesickness, which most nine-year-olds go through when they leave home for the first time. However, these periods did not last long. There was always more than enough to do and new friends to make. Even apart from the academic side it was a great learning experience, meeting and engaging with new people and new ideas.

Miss Long was the Form 1 mistress. We were a mixed class, as girls from the East Anglian School had been evacuated from Bury St Edmunds to Cadogan House, the prep department of Culford, the previous year. Although I could read and write, I had not yet learnt to do joined up writing. I was given a copy book and had to spend hours copying the scripts. During the term, we had to perform a little playlet about belling the cat. As mice we borrowed Brownies' uniforms for our costumes. I was the mouse that proposed we put a bell round the cat's neck but announced that it was someone else's job to sort out how it should be done.

Saturday mornings were devoted to writing a letter home to our parents, usually checked over by a member of staff. One boy, Josef, who was Czechoslovakian and had lived in Paris, would write in French. On one occasion, Dr Skinner himself checked over his letter and was very complimentary. Josef was the first boy I had met who could speak another language and I was most impressed.

When Father wrote letters to us at school, he would tell us what was going on at the farm and would often put in a little mathematical exercise. For example, *'if I was planting strawberry plants 4 feet apart on the square, how many would I need per acre? If I paid so much for picking blackcurrants and the crop yielded so many tons per acre, how much would it cost?'* I usually managed the puzzles but sometimes sought help. I never quite worked out how to deal with the field margins, a problem still with us 80 years later!

In the autumn term I was moved up to Mrs Farr's Form 2. We were a well-behaved lot generally and it only required a sharp word or two from her to bring us back into line. The following summer term, a new form, 2a, was created with

Miss Hare as our teacher. She was new to Culford and we tended to play her up but clearly some of the lessons made a lasting impression. I remember particularly that each week she would place six wildflowers in pots in the form room, beautifully labelled, so that by the end of term we had seen and had the opportunity of learning the names of most of the local wildflowers. On hot days she sometimes took us to have classes outside in the grounds of Culford Park. On one such occasion we were studying *The Jackdaw of Rheims,* a poem by Thomas Ingoldsby. Clive Young earned a severe reprimand when, fooling around, he tried to perch on a chair acting as the jackdaw in the poem.

Miss David was an old hand so, when we moved on to Form 3, we were kept on a tighter reign. The classes were held in a room off the quad at Cadogan House. We were continually moving from there to Main School where we slept, took our meals, and did prep. Sometimes books were left inadvertently in the wrong place. Dr Skinner raged into the form room one day wanting to know why some of us did not have the appropriate books for the lesson. The unlucky ones were sent to Main School to wait outside his study. After a while, as we pondered our fate, the 'Old Man', as he was known, though not to his face, returned. He ushered us in one by one and gave us each four of the best and admonished us not to let this happen again. Today I suppose this would be considered brutal or sadistic, but we took it as a matter of course and there was no animosity on either side. If one remembers that at that time so many were fighting and dying for the country, what was the odd caning?

I received a caning on two other occasions, including once in my pyjamas for being out of bed after lights out. This may partly have been because bedtime was 8.30pm but as it was double summertime it did not get dark until hours later. The caning on that occasion was given by Bill Blamey who had his residence in Edwards' Wing and would not stand any nonsense. He was known as Sarky Bill, and when we moved to Main School, he became our maths master. He was a brilliant mathematician, able to impart the various concepts of geometry, trigonometry, algebra and calculus with clarity and enthusiasm, but we did spend virtually one whole lesson having the difference between sarcasm and irony explained to us. The other misdemeanour for which I was chastised was for taking toast, butter and marmalade from the masters' breakfast table after everyone else had left the dining hall. I thought it was a pity to see it go to waste, but that was not what the staff felt about their ration of butter and marmalade, even if I might have been able to get away with eating the toast.

Apart from our form mistresses, we had other teachers too. Miss Johnson, the girls' headmistress, took us for one lesson a week, English Literature, and more specifically long narrative poems. I don't remember being particularly enthusiastic about it at the time, but she must have made an impression for I can still recall her taking us through the poem of *Sohrab and Rustam* evoking the wide-open spaces of the Asian Steppes. For geography we had Miss Hunter, another experienced and capable teacher.

The next year in 4b we widened our curriculum to take in Latin and French. Latin teachers were in short supply, so it fell to the Reverend Peter Morley, the school padre, not only to teach us RE but also to guide us through the cases, genders and ablative absolutes of Latin grammar. He was quite open about the fact that he was only two lessons in front of us. Michael Keall, a new boy who had been a choral scholar at King's College School, Cambridge, where he had already learnt some Latin, frequently crossed swords with Peter, though then we never dared call him by his Christian name, over the correct endings. Michael, known as Tom to us at the time, became a lifelong friend. He went on to be headmaster of Westminster Abbey Choir School, a governor of Culford and bursar of Magdalene College, Cambridge. Peter Morley was able to make religion immediate and fun, for instance when he likened 'speaking in tongues' to 'having the gift of the gab'. He was one who loved his Christianity and inspired others as much by what he did as by what he said.

As well as being a popular master at the school, Peter was also troop leader of the Scouts and led us in our many scouting activities: wide games, badges, camping. I had joined the Cubs in my first term and later moved on to the Scouts where we learnt all sorts of things as we worked towards achievement badges: semaphore, Morse code, first aid, knotting, cooking at summer camp, sailing on the Broads at Easter and a whole range of adventures which suited me down to the ground. About 500 yards to the north of Main School was the North Wood, and each summer for one weekend we would make camp on the edge of the wood. Each patrol would sleep in a bell tent, cook their own meals on an open fire and come together with the other patrols for some games in the afternoon. Just as it was getting dark, the evenings would be rounded off with a sing song and cocoa beside the campfire, led by Peter who had a good, strong voice.

Keith Bullock, the son of a neighbouring farmer was in the same year as me at Culford and we did a lot of things together. It was a friendship which continued through the years[3]. In the summer of 1947 about 40 of us Scouts and a couple of scoutmasters travelled by train overnight from Bury St Edmunds to Edinburgh. We took with us our tents, dixies[4] and all the paraphernalia associated with going to camp, together with a trek cart. It was a rather disturbed night as there were many ex-servicemen on the train and so we had difficulty finding somewhere to sit or sleep. We arrived at Edinburgh early in the morning, rubbing our eyes and ate our breakfast in town before boarding a bus to take us to Innerleithen. There we pitched our bell tents on the lawn of Pirn House, a dilapidated old manor house on the banks of the Tweed. After lunch most of the troop were able to go exploring, scrambling through the Tweed or climbing one of the nearby hills. I was rather put out as I was detailed to stay in camp to keep an eye on things. It may have been an

---

[3] Keith died in August 2021.
[4] Dixies: cooking pots with lids and steel handles.

honour to be selected for this, but I did not see it that way at the time. I did of course have my opportunities later.

The chance came to do our Hikers' badge, which Keith and I did together. We were given a map and told to find our way to a particular spot to camp that night. The map led us to the headwaters of the Leithen river where we pitched our small tent, cooked our evening meal, gathered some bracken to make a comfortable mattress, and spent the night. The next morning, we were greeted with a Scotch mist[5] as we made breakfast, struck camp and packed our kitbags, but as we walked through the damp heather down the hill and back towards Peebles, the sun came through, lighting up the town. It is a view I have remembered ever since, and I could not but think of the celestial city in *The Pilgrim's Progress*.

One night was given over to a game for which we divided into two teams. The object was to find the other team's headquarters where rockets were stored and to let them off. Wandering about in the dead of night we must have trespassed on many local properties but the father of our scoutmaster, Ian Nicholson, lived there and had pre-warned his neighbours accordingly. Another day we built rafts on the Tweed and on another were taken to Dunfermline where we had our first experience of a coal mine. Altogether I had a great time and loved the outside adventures, which is perhaps why I remember the trip so well.

Although the school was nominally Methodist, almost half of us were Church of England. Accordingly, there were two services on a Sunday morning. The C of E service took place in St Mary's church in Culford Park where each Sunday we had to go through the *Venite* and the *Te Deum*. Not surprisingly we sometimes opted to go to the Methodist service instead, which not only had brighter hymns but also was a shorter service. However, together with others, I was prepared for confirmation by the Reverend Sandford,

---

[5] Scotch mist: heavy drizzle.

the vicar. Looking though papers from my school days recently, I found the card given to me on the day of my confirmation, 27 March 1946, in St Edmundsbury Cathedral. Coincidentally, 27 March is my wife's birthday though I had yet to meet her. I got to know Reverend Sandford rather better three years later as I was churchwarden in my last year at Culford, responsible for counting all the threepenny bits with which we were issued to put in the collection bag. Little did I know then that only 10 years or so later I would become churchwarden of St Mary's Bacton, a post I was to hold for over 50 years.

Having been brought up on a farm, weather was a daily topic of discussion at home, so perhaps it is no surprise that some memories from my school days relate to extreme weather events. The summer of 1946 was very wet with some severe hailstorms. At school, the glass of the rotunda on Main School was so damaged by hail that the rain poured in during a storm. There was also much damage on the farms in Suffolk and I remember anxiously waiting for my weekly letter from Father to know how they had fared back home. Overall, we did not suffer too much but there was a strip about five miles wide and 30 miles long through mid-Suffolk where the hail stripped the leaves from the sugar beet and wheat and barley was beaten to the ground. The following year, 1947, was the year of the great freeze. It started in the middle of January, a couple of weeks after we had returned from the Christmas holidays, and persisted until the middle of March. The lake in the school grounds froze over and we went sliding or skating most days; no field hockey, but we did our best on the ice. It was jolly cold in the Main School hall and I remember sitting reading on a radiator trying to keep warm.

I studied chemistry, physics and biology in the sixth form, gaining my Higher School Certificate, which today would be called A levels, in these subjects. There was talk of university from the headmaster but it was always expected that I would return to the farm and make my living there. I have no regrets.

I enjoyed and participated in all sports, though I was never much of a cricketer. I was enthusiastic about tennis, gymnastics, athletics, hockey and rugby and played in the school teams at all levels. I remember both parents coming to watch me play rugby in the autumn of 1946, bringing with them, in a carry cot, my new brother, Thomas, who had been born on 26 September.

In my final year I was captain of the hockey and rugby first teams when we achieved middling success against the various other schools. It was in the summer term of that year, 1949, that I mopped up many of the athletic events, winning the cross-country, the mile, half-mile, 440 yards, 110-yard hurdles, high jump and shot put, though not, I must admit, all on the same day. My earlier misdemeanours which led to the beatings could not have counted too much against me for I was also school captain, dishing out punishments rather than receiving them.

That summer, when term ended, brother Robert and I joined a school trip to Switzerland, our first trip abroad and an eye-opener. We travelled by train to Geneva and on to Montreux. We visited the sights: the chateau of Chillon, les Dents du Midi, the Great St Bernard Pass. At the pass we walked across the border into Italy, which seemed a wonderful thing to be able to do. We bought ice creams but had to pay with our Swiss francs. To our amazement, the change we were given, worth only a few pence, consisted of bundles of lira in rubber bands, such was the inflation rate in Italy. We visited alpine farms where we sampled the cheeses and saw the Swiss brown cows grazing with tinkling bells around their necks, quite a different scene from Suffolk. We marvelled at the contrast between austerity England and the bright lights around the shores of Lake Geneva. No food rationing there; ice creams of all flavours were served from the kiosks on the brightly lit promenade. Food and sweets, still rationed in England, were freely available. We even brought home some icing sugar and desiccated coconut for our grandmas, who were keen cooks. This trip brought my school days to an end on a high note.

Culford Hall, West Front

With my brother Robert at Culford School c.1942

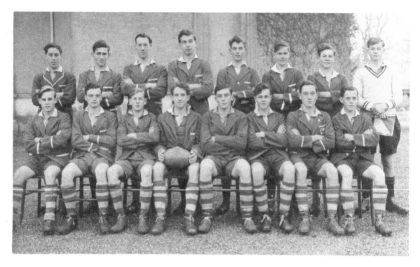

Culford School 1st XV 1947
Back row L-R: Housden, Keith Bullock, unknown, unknown,
Geoff Doubleday, David Black, Bryant, Clive Young (linesman)
Front row L-R: Ian Marriss, Duncan Drummond, Burgess,
Ian Hendry, Caruth, unknown, unknown, unknown

Culford School 1st XI, 1949
Back row L-R: John Hawtin, Bryant, Geoff Doubleday,
Burlingham, unknown, Housden
Front row L-R: Trower, Peter Stockley, David Plastow,
David Black, Bill Belcher

Taking part in Sports Day, Culford School, 1949

Victor Ludorum, Culford School Sports, 1949

Prefects, Culford School, 1949
Back row L-R: David Keate, David Plastow, John Hearnshaw,
Peter Stockley, Pavitt,
Front row L-R: Richard Freeman, David Black,
Michael Keale, Ronnie Moore

My final School Report, July 1949

CHAPTER 3

BIB AND BRACES

Having left school, I now had to work in earnest alongside my father and grandfather, slowly learning more about the farm as a business. By 1949 we owned and farmed about 1,200 acres employing 50 men and 15 women full time. I shall try to give an idea of the range of activities that I was involved in on the farm, but I make no apologies if the nature of some of my recollections makes the whole seem rather disjointed.

There was no formal programme to initiate me or to ensure that I experienced every aspect of farming. The same applied to my brothers, Robert and Thomas, when they joined the farm later, Robert in 1951 and Thomas over a decade after. Grandad said he could teach us more on the farm than we would learn if we went away to agricultural college, so we never went. He argued that if we were to learn from our mistakes, the sooner we started the better, although he pointed out that it would be cheaper and less painful to learn from the mistakes of others. So we learnt continually through our work on the farm and through the Young Farmers' Club (YFC).

Father was always very mechanically minded and ever since I can remember we had a workshop on the farm, with a lathe, a drill and a grindstone. However, it seemed to me that I could learn rather more, machinery-wise, off the farm than on it, which is how I came to spend a few weeks in early 1950 with Cornish and Lloyds of Bury St Edmunds, a firm of agricultural engineers. Mother packed my lunch and I commuted there in an old Morris van. I had hoped

I might have some instruction in lathe work, for they had a machine shop, but they were basically too busy to spend time trying to teach me. However, I observed and picked up what I could. Iron casting work was done once a week and after the moulds had been prepared I watched, fascinated, as the molten metal was poured. Cornish and Lloyd made a variety of different castings including pig troughs, pulleys, table legs and name plates. When these castings had hardened and cooled sufficiently, the mould was broken open and the rough edges were ground off, while the pulleys were put in the lathe to be finished. I was allocated one job to do on my own. After being shown how to tighten brass bearings by filing I was told to get on with sorting out those on the connecting rods which had come, rather appropriately I thought, from the deep well pump at Culford School, Cornish and Lloyd having the contract to refurbish it. My last job there was to check over, though under tighter supervision, an International Farmall M tractor which Father had bought, before it was delivered to Red House Farm.

Back on the farm, my role was to help with whatever job needed doing or whatever Father or Grandfather thought was the most important task at the time. Cutting flowers, picking apples, stacking straw, tractor driving, feeding pigs, mending machinery and lorry driving, I had a go at them all. Given the variety of crops growing on the farm, the tasks were ever changing. It was difficult to keep up with everything but, as time went on, I gravitated towards the arable, pig and building aspects of the business.

Grandad always returned from his various meetings with ideas about growing new crops. He expected Father to try them all out on the farm which often led to some heated arguments, but consequently we became involved in a wide range of crops. We were growing not just the usual Suffolk crops of wheat, barley, clover and sugar beet but Grandad's passion for trying something new also had the business growing fruit: apples, pears, plums, blackcurrants; cut flowers for the London market – daffodils, tulips, trollius, iris, paeonies and gladioli; seeds for McGill & Smith, cocksfoot, timothy, tall fescue and red and white clover; vegetables including potatoes, cabbage and sprouts, mostly for the Co-op in Ipswich to put on their electric trucks to deliver to retailers in the town, as well as rhubarb, special stock, virus-free strawberry runners, peas and mangolds. The wheat and barley were mostly ground into meal to feed our own pigs, though occasionally we could produce a barley sample good enough for malting. You name it, we seemed to grow it. During the War we even grew poppies for the seed to be processed into morphine.

The day always started early on the farm, and it is a habit so ingrained in me that I still wake to *Farming Today* on Radio 4 at 5.45am each weekday. Mother would be up at 6am preparing breakfast and, by 7 o'clock Father and I would have seen the day's mail and be ready to meet the men outside the farm office to give instructions for the day's work. Grandad would be up and about early too, though he would pop back indoors later for breakfast.

Grandad mostly attended to the horticultural side of the business, overseeing the work on the fruit, flowers and bulbs, with Bill Alexander as his foreman and Les Forsdyke as his number two. There was also a gang of ladies who would be doing whatever work was needed depending on the time of the year: pulling rhubarb, hoeing and weeding, cutting and packing flowers, lifting bulbs, cutting asparagus, picking fruit and so on. It was up to Father to keep everything running on the arable side, for we also had land at Wickham Skeith, Norton and Mendlesham, but when more help was needed with the flowers or fruit then Grandad would 'borrow' some of Father's men, making his job more challenging. Not infrequently there were minor arguments as to what was most important. Father told me he had given up allocating tasks and instructing the workforce the night before as he often had to change the plans because of the weather, the markets, or even

Grandad's whims. In due course, I was expected to pass on instructions to the men. I suppose, as Father's son, I had a certain authority over the men working on the farm, but I was always careful when I wanted something done to ask what their previous instruction had been from Father or from Grandfather, for these were sometimes at variance.

Our land reached the giddy heights of 68 metres (223 feet) above sea level at High Pastures, falling to 60 metres (197 feet) at Red House. My father would point out that High Pastures was the watershed, with water from here draining either westwards to join streams heading north to the Great Ouse and thence out into the Wash, eastwards to join up with the River Waveney, entering the North Sea at Lowestoft, or south-east to meet the River Gipping, which flowed into the Orwell and out to sea at Felixstowe.

All our land was heavy boulder clay and needed under-draining, for without it, in a wet time, it would be completely unworkable. Even having drained a field, if the pipes were too far apart, a tractor could still get stuck in the wet clay. So it was that one of my early roles on the farm was putting in drainpipes alongside Charlie Quinton, our chief draining man from the 1920s to the 1960s.

I spent a month with Charlie and Kenny Betts working in a field at Potters Farm, Mendlesham, where we needed to put in extra drains. On good sloping land it was little trouble to get a fall, but on our flatter land it could be a challenge, one which Charlie cheerfully took up and took great pride in being able to do well. However, we would sometimes have a water cart beside the drain so we could test the gradient to be sure that the water would run in the direction we wanted it to. The trenches for the drains were originally laboriously dug by hand to a depth of 30 inches or more but Father was always trying to find ways to reduce the amount to be hand dug. The simplest way was to plough out a furrow as deep as possible and then to dig into that.

He built several machines pulled with a Caterpillar D4 which would take out a channel up to 20 inches deep, but those last inches still needed hand work. We later bought a rotary-hoe drainer fitted to a Fordson tractor, one of the early models brought in from Australia. This would cut a trench to the level we needed, but the early models hardly had enough horsepower to cut a level trench every time, particularly when they hit some particularly hard clay spots, so some hand work was still needed. Kenny Betts used to drive the rotary-hoe drainer, and he and Charlie together drained many acres.

Once the drain had been dug and levelled, 12 inch-long clay pipes were laid end to end in the bottom. We used 2 ½ inch diameter pipe for the laterals, leading into 3 inch and 4 inch pipes, and sometimes 6 inch pipes, to discharge into the ditch, depending on the area to be drained. The pipes we used had circular holes but in previous times all sorts of different shapes and sizes had been used: oval ones, some with a flared end, others with a flat tile on the base and a half round above. At one time we had a collection of a dozen or so picked up from different fields.

In the 1950s and 1960s we used the prunings from our blackcurrant bushes, tied into bundles and then trampled down on top of the drainpipes, to help water percolate through more easily. Nowadays, shingle is used for the same purpose. Water flowed through these sub-soil drains into the ditches, which needed to be cleaned out periodically to keep the water flowing and, until we bought a McConnell ditcher, this was another job which had to be done by hand. The McConnell machine fitted on the back of a Fordson Major tractor and Derek Steed spent a month or so each winter cleaning out ditches with this set-up.

Ploughing out first spit for a drain, 1943

Draining, 1950. Charlie Quinton in foreground

Me on rotary hoe drainer, 1956

Clay drainpipes outside Somerville, 1950s

We were fortunate that the government could see that effective drainage was necessary for good crops. For a period immediately after the War when it was critical to produce more food from our farms, it subsidised the operation by up to half of the cost. Each county had an agricultural drainage officer and the work had to be inspected and mapped to his satisfaction before the grant was paid. East Suffolk and West Suffolk were, at that time, separate counties, so with our land spanning the boundary, we worked with two such county agricultural drainage officers. However, over time, the subsidy was reduced and eventually removed altogether. Subsequently, with higher-horsepower tractors and the advent of lasers, the whole drainage operation has been revolutionised. It is now possible to employ machines which can cut a trench to the required gradient and lay perforated plastic pipe fed directly off a reel into the trench.

After drainage, the second priority for us on newly purchased parcels of land was the removal of hedges to make the land into bigger blocks. A typical such project was College Farm which we bought in 1953 adjoining Red House Farm at Earls Green, Bacton. It was a farm of 69 acres and had 13 separate enclosures, making an average field size of about five acres. For an arable operation this made no economic sense. The appropriate field size has been going up as the size of the machinery has increased. Although our fields in the 21st century may now be as big as 100 acres (40 hectares) we have over the past few decades been developing wider field margins and planting hedges in more convenient places to replace some of those that we had previously done away with.

Access was another requirement. The original tracks, following the old field boundaries, zig-zagged and were often impassable in wet weather. Gateways with their gates and gateposts needed constant attention if they were to perform their original purpose, but with no grazing livestock they were no longer needed and it made sense that, where we were putting down hard stone or concrete roadways, the roadways should be straight.

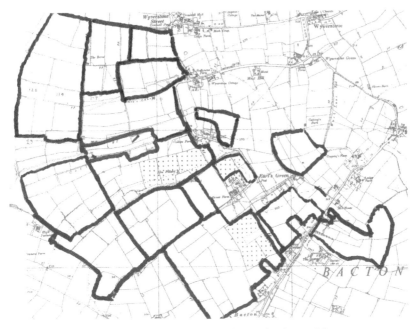

Map showing old field boundaries with
with enlarged fields marked in bold

New hedge, typical of 1990s

After harvest in 1948 we set about improving the access to High Pastures, 169 acres of land joining onto the Red House land at its western extremity, which we had bought in 1936. I remember that Father set up a stick in the High Pastures fields and told Kenny Betts to bulldoze a track to it straight from the end of Canada Lane. The track was filled in with stones from the Badwell pit giving us an all-weather roadway to High Pastures. Grandad was none too complimentary about the process but in the succeeding years he gave it his grudging approval.

At that time, the Avenue, or the Walk as we now know it, was only a grass meadow into which the horses were turned at night. There, another track was bulldozed out and filled with stone. This proved a popular route to the farm, a shortcut to Stowmarket compared to the Earls Green entrance which we had used until then. Later, the surface was tarred several times and it is now the main access to the farm. The Avenue justified its name because it had a magnificent avenue of elm trees. These unfortunately succumbed to Dutch elm disease in the 1950s and had to be cut down. Some 20 years later, brother Tom had a strip of mixed trees planted on the west side of the road which now presents an attractive visual break as well as a wildlife haven and a replacement for some of the hedgerows we have removed elsewhere.

Grandad travelled frequently to London on NFU business, taking the train from Finningham station. However, there were no parking facilities there and so I often acted as his chauffeur, running him to the station in the morning and collecting him again in the evening. I would sometimes wait in the signal box for the train to arrive, chatting to the signalman, Freddie Cuthbert. He had levers in the box to operate the signals, indicating whether the trains could continue or had to stop. Occasionally he would let me pull a lever to change one of the signals, but being connected with rods, it took a great deal of effort. Incidentally, years later, Freddie's grandson, Andrew Cuthbert, came to work on the farm straight from school. Andrew worked alongside Les Gooderham in the workshop for many years before becoming the farm's main mechanic, and stayed with us right through to his retirement in 2020.

On another occasion as chauffeur, I remember driving Grandad home from a farmers' meeting at Holt, Norfolk late one night. We were on the straight piece of road which runs through Thetford Forest when he said to me, 'Come on boy, do you think we could do 100 miles an hour?' We did.

We always seemed to be repairing some existing machinery or making something new in the workshop and employed a full-time mechanic. When I started work on the farm, Nathan Garrard was the mechanic running the workshop. During World War I he had worked in Mesopotamia (now Iraq) building and erecting telephone masts. He lived in one of our farm cottages and was a keen gardener. From the new varieties often seen flowering in his garden, which were strangely like those Grandad had bought, it appeared that Nathan was not averse to taking the odd bulb home. We had a carpenter, too, Cecil Gaffer, and often Kenny Betts or Tiny Ritchie would help in the workshop when they were not needed in the fields.

Horses were never my strong point as is evident from my early encounters, but we still had six or seven in the stable. Although their importance was waning, until the late 1940s horses were still being used for drilling, rolling, hoeing beet, carting muck, sugar beet or mangolds, and ploughing out daffodil and tulip bulbs or gladioli corms. Grandad said we would never be able to cope without at least one horse on the farm, but the little grey Ferguson (Fergie) tractor that didn't need feeding, grooming or putting out at night and rounding up again in the morning eventually sounded their death knell. We continued to use horses for mucking out work in the pig sheds some time after we had stopped using the horses on the land. One could say to the horse, 'gid up' and the horse would move forward, whereas with a tractor, you had to get back on to move it. Our last horse went in 1950 but we kept the pony longer. Bert Churchyard, the third horseman, always known as Squiggler, though we never did find out why, used the pony with a flat-bottomed cart to bring in the apples, plum, and gladioli, and for all manner of little carting jobs about the farm.

Sugar beet drilled in this way until 1950.
Smythe drill with horse and two men

Rolling beet

Squiggler with pony, carting potatoes, 1950s

With my Grandfather David Black by
the horse meadow beside Somerville, 1955

Planting potatoes c. 1960. L-R: Teddy Cook,
Little George, George Draper

Picking potatoes, 1950s

Riddling potatoes from outdoor clamp, 1950s

Frank Johnson, Kenny Betts and
Monty Mutimer cycling to work, 1962

William (Butcher) Betts had started work as a bacc'us boy for Grandma, before World War I doing odd jobs around the farmhouse such as chopping firewood, fetching water, feeding the chickens and guinea fowl, and collecting the eggs. Later, he was given the job of driving our first lorry. It was used for carting food for the pigs from Ipswich docks, taking pigs to the bacon factory at Elmswell, collecting fertiliser from the station and all the other fetching and carrying jobs that needed doing. I understand that with this first lorry on the way home one day, he had to stop at the traffic lights in the middle of Stowmarket. He was on a hill and, when the lights turned green, he let the clutch in too quickly, the front wheels came off the ground and a sack or two fell out of the back. Apparently, it took some time to master those early clutches. We had a succession of lorries, each an improvement on the previous one. Father insisted the lorry bucks be painted green and the cabs red so that they could be seen better and would be safer on the road. Despite that, I recall Butcher having a collision with another vehicle at Stowlangtoft Spinney crossroads. At the inquiry, when he was asked if he had looked left and right, he famously replied that he could 'see without looking'.

Our first tractors all had starting handles, with no electrics or self-starters. They ran on TVO (tractor vapourising oil) or paraffin as we called it. The engine had to be hot enough to vapourise the fuel so first the engine had to be started on petrol and then, after a few minutes, it could be changed over to paraffin. Patience was called for and woe betide if you ran out of petrol and tried to start on TVO. It just would not go. If we were taking fuel out to the tractors in the fields, which we often did in five-gallon drums, care was needed to ensure there was no confusion between the paraffin and the petrol. We became adept at being able to smell the difference, though not before Robert and I inadvertently put paraffin in a petrol tank on a baler and spent an abortive couple of hours trying to start it. Fortunately, Father came along, took a sniff, and told us of the error of our ways.

The first tractors seemed to need as much care and coaxing as the horses had done. Those tractors with steel wheels or tracks were usually left in the field overnight if there was still work to be done. There was no antifreeze so, when there was the suggestion of a frost, we had to drain the water from any engine at risk before we could leave the field at the end of the day. Usually there was one drain tap at the bottom of the radiator and another on the engine block though sometimes these would block up and need to be cleared out with a piece of wire. Of course, on a frosty morning it would not do to put cold water into the radiator and so we had to take hot water with us to the field as well as fuel for the day. Antifreeze, self-starters and diesel engines were a great step forward.

We grew sugar beet for both the Ipswich and the Bury St Edmunds factories. At the beginning of the season, in conjunction with the beet fieldsman from the factory we had to forecast how much beet we would produce. We were then issued with permits to deliver the beet over the period of the sugar processing campaign which usually lasted from mid-September to mid-January. I had one season carting sugar beet in the lorry. The beet would be pulled, knocked and topped by hand, then carted to a heap beside the road. When we needed to send a load, then about seven tons, we would call in help from the beet field, and fork it all on by hand, a laborious task. It also had to be forked off at the factory which was hard work on one's own. Tipping lorries, when they came, were a great boon, speeding up this job and easing the work of the lorry driver.

The lorry cabs were unheated so at least loading the lorry meant one started the journey warm. Sometimes we would load the lorries the night before and, when other farmers had done the same, one would see lorries queued back from the beet factory to the Angel Hill in Bury waiting for the factory to open at seven o'clock in the morning. If the loaded lorries were left out overnight and there was a sharp frost the load could freeze solid to the buck. It was not

Cutting lucerne, May 1956

Loading lucerne onto trailer to take to
grass drier in Top Shed, May 1956

Cocksfoot seed for McGill and Smith, 1961

Charlie Hales ploughing with triple-D, 1958. Machinery dealer,
Ernie Doe, devised this solution to provide four-wheel drive and
double the horsepower by linking two Fordson tractors together.
In a wet time on our heavy land, four-wheel drive had a decided
advantage over the traditional two-wheel drive tractor

unknown for a lorry tipping beet into one of the silos to topple backwards with the load. Beet could also get caught up on one side of the buck and if the driver did not let it down quickly enough, the lorry could tip over sideways.

We were always seeking improvements and sometime after my first season carting beet we bought an elevator to load the beet although the beet still had to be forked into the elevator. It had a flat belt drive from a one and half horsepower petrol Lister engine. We had to remember to take petrol to the field for the elevator as well as for the tractors. It certainly made the job easier, but there is no way, with its unguarded belt, that it would pass today's safety regulations. We made extension sides for the lorry buck so that we could carry more beet, greedy boards we called them, but even so the maximum load we could carry was only seven tons. In the early days of sugar beet growing, we were only averaging nine tons per acre in Suffolk. Nowadays, in a good year with improved varieties and land management including weed control, we can achieve 40 tonnes per acre.

Over the years, lorries became bigger, capacities increased, and braking capabilities improved. During the 1950s, four-wheeled lorries were allowed a maximum speed limit of 30mph while six-wheelers, capable of carrying more, were limited to 20mph and were obliged to have a driver's mate. Occasionally, in a hard frost, a wax would settle out from the diesel, blocking the fuel pipes, though later we found we could put in an additive to prevent this happening. The other possible blockage was when the water in a sediment bowl froze. This bowl was there to collect any water which may have condensed in the fuel tank. There was much to be said for parking the lorries inside when we had frosty weather, but it was not always possible.

With the amount of manual handling to be done on a farm in the 1940s and 1950s a good level of fitness was required. Today it is compulsory for manual workers to attend training courses before they can lift heavy loads and there are restrictions on how heavy a load can be. However, in this era before mechanisation, we did the best we could and were proud of what we could lift and carry. As far as I know no one on the farm suffered from more than aches and pulled muscles. A regular lorry job was to take the meal in hessian sacks from the mill to the various pig units. We slid the sacks down a plank from the first floor of the mill onto the lorry for carting to the pig sheds. We also bought and collected wheat from neighbouring farmers to mill for pig feed. Initially we handled wheat in 'coomb' sacks, each weighing 18 stone or 252 pounds, and it was a great improvement when we had a hydraulic sack lift fitted to the back of the lorry. It was easy to wheel a sack on a sack barrow to the sack lift, which then took the sack up to shoulder height. Once there, you could put it across your shoulders, carry it to the front of the lorry and drop it into place. I have carried hundreds of tons that way. The hessian grain sacks were used for all sorts of other purposes: as aprons, as cloaks to keep the rain off, as mats. It was one of our standard winter jobs, often with snow on the ground, to check through the sacks and mend any with holes, often caused by rats or mice.

We collected other feeding stuffs for the pigs including middlings[1] from the local millers, Cranfields, at Ipswich docks and from Marstons at Icklingham. For two or three years we used large quantities of tapioca meal which came into the country on small lighters to Wivenhoe on the River Colne having been transhipped in Rotterdam. This was a very dusty crop and the streets in Wivenhoe became covered in a white dust. Eventually the residents complained bitterly, and the importation was banned.

It was in 1950 that my father decided we should build some silos for grain storage. Initially he was attracted to some circular silos made up of concrete sections. We bought enough sections to build two silos approximately 14 feet in diameter and 15 feet high. We laid a concrete pad next to the

---

[1] Middlings: the part of the wheat grain between the outer bran and inner endosperm.

grinding shed and started to erect the first one. The next morning, he came out and said 'Stop'. He had obviously been worrying all night. 'Round silos can't be the right answer,' he said. 'There will be too much wasted space between them, space for rubbish to collect, possible problems with rats and mice. Let's build some square ones.' Jack Aves was building houses for us at the time, and Father went to consult him. The upshot was that we delayed building any silos until the following year, when we had Aves build a block of eight, 15ft sq silos in brick, four either side of a central tunnel, each capable of holding 80 tons of grain.

Elevators and belt conveyers were needed. We had already made elevators for handling meal and so Father decided we should save money by making our own. I spent much of the following months helping to make and install the appropriate equipment. It was quite a challenge as we had little in the way of drawings to go by and had to make up much as we went along. Even then, it could be quite frustrating as Father was apt to change his mind as to exactly where elevators and chutes should go. Eventually we got them all made, installed and going and at the same time installed a Penny and Porter dryer to ensure we could dry the corn before storing it. In the end I am sure that Father made the right decision with the square silos, but there was little experience and what advice there was sometimes pulled us in different directions.

Yet more silos were added in 1960 and again we did much of the work ourselves. This time however we had some drawings to go by and we used shuttered, reinforced concrete instead of bricks. I was master-of-works in all but name and learnt as we went along. We started the foundations in March and made enough wooden shuttering to encompass all eight silos to a height of 3 feet. We mixed the concrete using a fore-end loader, one shovel of sand to three shovels of shingle and one bag of cement, then wheeled it to the site in a wheelbarrow, shovelled it into the shuttering and vibrated it to exclude any air pockets. Three days later, the concrete having set sufficiently, the shuttering was dismantled and re-erected up a stage together with the reinforcing ready to be filled again. As we proceeded, we had to erect scaffolding and lift the wheelbarrows onto the decking using another fore-end loader. Freddie Whatling (the son of Grandad's head horseman), Bob Betts, Dick Alexander, Frank Johnson and Tiny Ritchie formed the gang working with me. Looking back, I wonder how we did it, for we started at 7am, had half an hour off at 8.30, another half hour at 12, then a break for tea from 5.30 until 6pm, after which we put in another three hours' work before breaking off for the night. We made a good, cheerful gang but were pleased when we got to the height when Aves' carpenters could come onto site and fix the roof.

In my early days on the farm, fertilisers were usually delivered to and collected by us from the railway station at Stowmarket. The fertiliser came in sacks each weighing 1cwt[2] and these had to be hand loaded. It must have happened many times, but I remember on one particular day being detailed to move 120 tons of fertiliser from the station to our different farms. We had three lorries doing round trips but, since each lorry could only carry seven tons, we had to cope with 17 lorry loads. Although there was a gang of four loading and the same number unloading, we were more than a little tired by the end of that day. Some fertilisers could set solid in the sack over time, making them awkward as well as heavy to handle, and then we would have to spend hours breaking down and mixing super phosphate with sulphate of ammonia and Kainit (a potash fertiliser) before spreading onto the various crops. It was a great boon when free-flowing, granulated fertiliser appeared on the scene.

Another lorry job I remember was being sent to collect some sprays from Fisons Pest Control: arsenate of lead, nicotine, lime sulphur and winter wash, for use on our apple trees. There may have been a few concerns raised from time

---

[2] Cwt: a hundred weight, equivalent to 112lbs or approximately 50kg. There were 20cwt in a ton therefore 2,240lbs. Today a metric tonne is 1,000kg, equivalent to 2,200lbs.

Les Forsdyke spraying apple trees, 1950s

to time about the suitability of the use of these products on fruit trees, but it was generally accepted that they were necessary for the control of pests and diseases. Selective weed killers for use on wheat and barley were just becoming available. MCPA was one of the first and its effect on charlock[3] was amazing as at last we were able to control one of our most competitive weeds. I collected the first crop sprayer we ever had on the farm, a 60-gallon machine which fitted on the back of the Ferguson. The nozzles were forever getting blocked, for at that stage we hadn't sorted out the filtration as well as we did later. At the time it seemed as though a silver bullet had been found, sprays providing nothing but benefit on the farm. Since then, we have learnt much more about the effects of these chemicals on the bugs, weeds and diseases which we set out to control, on the personnel using them and on the environment. Now aware of

---

[3] Charlock (Sinapis Avensis): an annual weed which can cause damage to arable crops. It produces many seeds and spreads rapidly.

the potential deleterious effects that some chemicals can have, some have been banned completely while others have been put through a vigorous testing regime which was unthought of 60 years ago. Sometimes we think the pendulum has swung too far in the other direction as the 'precautionary principle' has been invoked in a way that stops developments in what could be beneficial directions. Whatever, the use of all chemicals is now under strictly controlled conditions.

One of the jobs which I did for four or five years was helping to deliver strawberry plants into the Wisbech/Kings Lynn area. We grew special virus-free plants. The runners would be lifted in October and a gang of girls (for so we called them, though maybe ladies or women would be more politically correct today though even that seems to be challenged by today's woke society), tied them into bundles of 25 plants ready for delivery. We sold them to a hundred or so part-time smallholders who grew strawberries for the Wisbech jam makers. It was worth their while to buy new strawberry plants every two or three years because the old plants would succumb to various viral diseases which meant they would not fruit well. There were many deliveries to make; each lorry load would have around 100,000 plants and we had to drop off 500 plants here, 1,000 plants there. It was a full day's work for two, allowing us to alternate the driving. Depending on who was available, my co-driver could be either Horace Bloom, Tiny Ritchie or brother Robert. Mother would always pack a Thermos flask of hot tea and sandwiches which we would eat when there was time. On some occasions, we were driving in thick fog which did not make the task any easier, and it could be a bit scary on those dyke-edged, Fenland roads. Some of the villages were several miles long for, as the Wash was reclaimed, the villages extended into the reclaimed land. Gedney Hill and Gedney Drove End were around 15 miles apart, so locating a particular smallholding in Gedney was not easy unless it was one which we had remembered from a previous year.

We grew cabbages, mainly January King. These were drilled in June or July and harvested when the market was favourable, sometimes in the snow and sometimes in that gap for home-grown fresh vegetables in March or April. It could be a cold job. I worked with the men to pack the cabbages into bags, sew them up and then load them onto lorries for delivery to Covent Garden, standing on them as we did so. There they would be thrown off, sold and reloaded for the wholesaler who then delivered them to the greengrocer. By today's standards, they looked in a poor way by the time they reached the consumer. The supermarkets have done much to improve the quality of what is on offer. In some years, the cabbages could be very profitable and in others we would have to plough them in. It all depended on the supply and demand, and I learnt early on that there were lean and profitable times in farming.

I still had much to learn about business practices and was perhaps á little naïve. I had always been brought up to believe that my word was my bond and did not necessarily need paperwork to confirm a deal. However, I was to be given a rude shock. We had supplied Chivers of Histon, Cambridgeshire with blackcurrants for their jam making from the time we had started growing the crop in the 1930s, Grandad having a verbal agreement with Stanley Chivers whom he knew quite well. However, Schweppes took over Chivers in 1959, and asked to see our contract for supplying them with blackcurrants. With no paperwork, Schweppes would not honour the terms of our verbal agreement and offered us a much lower price for the currants. It was unfortunate and I am much more wary nowadays and, as Grandad had always said, you learn by making mistakes.

Changes and developments were taking place in all areas of farming at what seemed to be an increasing rate. Father, Robert and I would visit other farms and go to the various shows to try to keep up with the latest ideas and equipment. The Suffolk Show locally, and the Smithfield Show held at Earls Court in London, gave us a chance to meet with various machinery makers, discuss developments and endeavour to sort out which would endure, and which would fall by the wayside. Some years later I travelled with my wife Mary to the French agricultural show held at the Porte de Versailles to find out what was happening on the continental scene. Each year there were bigger and more powerful machines available. However, we had to be careful not to rush into buying new machines until we knew that spare parts would be available back in the UK.

Hydraulics were developing fast, and it was a great step forward when we were able to load a trailer with muck using a fore-end loader on a tractor instead of a hand fork. The first loader we had was fitted onto our David Brown Cropmaster. I took it on one of our lorries to Jacksons of Bruisyard, agricultural engineers, who fitted the loader, and then delivered it to the Norton farm where we were mucking out the cattle yard. This early model made hard work of solid trampled muck but, step by step, the manufacturers have increased the loads that can be lifted and the height to which they can be lifted. A teleporter or front-end loader is now an essential tool on any farm and has done away with so much hard work.

It was not long before I took over some responsibility for the arable and pig sides of the business from Father, working closely with good staff, some of whom were with us for over 40 years. Most days I would spend some time in the office, planning the next building, checking up on prices for grain and other feedstuffs for the pigs or sourcing bricks as we expanded the pig business. I was in and out of Clarkes at Walsham frequently as we were buying so much building material from them. Often, when a repair in the workshop was being held up, I would dash over to Freddie Stennett's or Brian Legg's to pick up the odd bearing, oil seal or hydraulic fitting. I could be over at Norton catching up with Roy and Derek Steed who managed the crops there, or at Wickham dealing with a problem that had arisen, or looking at the crops in the fields, or fixing something in the mill. I was always keen to keep things moving and my overalls served a useful purpose as I continued my hands-on approach, working alongside the men whenever required. Sometimes I had to call in at home for a quick shower and to change into more suitable clothes before driving to a meeting at Fram Farmers or at the bacon factory. No two days were ever the same.

Sprout picking, Tommy Wright, 1950s

Bill Alexander weighing blackcurrants, 1960s

Tulip picking, 1961

Indoor rhubarb, 1967

Farm lorries carting sugar beet, late 1960s

On the farm with Father, 1960

On the farm with Robert and Father, 1960

In the farm office, 1970s

CHAPTER 4

THE YOUNG FARMERS' CLUB

I joined Stowmarket YFC shortly after leaving school. The movement had started between the wars, with youngsters looking after pet animals and hand-feeding calves on farms, and developed as they wanted to show them off and meet up with others doing the same thing. During World War II the YFC was seen as a means of encouraging food production and so was strongly supported by government and by local councils who provided funds to employ a county organiser. By the time I joined, there were a dozen YFCs in east Suffolk and the focus had widened. Each club ran its own affairs, strongly supported by the local farmers. There were around 30 of us in the Stowmarket YFC in 1949, mostly farmers' sons and daughters. In the winter we met weekly in the evening in one of the Stowmarket schools to hear a speaker, while in the summer we usually went on farm visits. I enjoyed the opportunity to meet up with others with similar interests, having some fun and a chat as well as picking up on the latest goings on in our day-to-day business.

During World War II most food was rationed and everything we produced on the farm had to be sold to the Ministry of Food; these regulations were slowly being eased off, but whether due to a shortage of home-grown wheat or because the country could not afford to buy it, bread was put on ration after the end of the War, and meat was to stay rationed till 1954. Not surprisingly the government was keen to encourage production by whatever means it could, and this fervour rubbed off on us.

The flavour of our meetings tended towards developing practices which would improve our outputs. Some were quite specific. A popular speaker was Reggie Flatman, fieldsman for one of the sugar beet factories, who, interspersed with a good deal of humour, gave us advice on working practices such as how to prepare a seed bed, and outlined his view of the pros and cons of various pieces of machinery. The local vet was another, talking about pets as well as farm animals. Salesmen and local farmers all came to educate, to enthuse and to enlighten us. One of our members told us of his experiences of one-way ploughing and its benefits. This was where one could start ploughing on one side of the field and steadily progress across the field, doing away with the hassle of setting up tops, making sure they were parallel and leaving a furrow every so often. Although it had the added expense of having to have left- and right-handed shares and breasts, this is the way ploughing has changed and nobody would want to go back to the previous method. Interestingly, with concern now about global warming and an emphasis on reducing carbon footprints, the concept of ploughing itself is being questioned with the possibilities of occasional no-till and mini-till systems being extensively explored but which I believe would be impracticable without the use of chemicals such as Roundup.

We also heard about new sprays being developed to control weeds, pests and diseases, about better use of fertiliser and how tractors and farm machinery could move on from just being replacements for horses and horse-drawn implements. We learnt about hydraulics coming onto the scene, enabling the development of fore-end loaders and of how bulk handling could replace the use of sacks and bags. It was a great step forward when we no longer had to go out after tea to collect the 18 stone sacks of wheat left by the combine.

In the summer, most farmers were keen to show us round their enterprises and to share their knowledge and ideas. It was easier then to look round livestock units than it is today, as biosecurity was not the issue it is now. I was rather embarrassed on one occasion when I had organised an evening meeting which coincided with a dry spell of weather after a week of rain. Only three members turned up at Bill Johnston's Sunnyside Farm, the others making the most of the good weather to get their hay in. Mr Johnston had gone to the trouble of getting two or three representatives from ICI, manufacturers of fertilisers and spray chemicals with a division specialising in giving advice to farmers, to help show us round. There were more people to do the showing than there were Young Farmers to be shown. However, he was very understanding when I, in my role as chairman, called by the next day to apologise and to thank him.

Weekend residential courses were arranged on a county basis. In Suffolk, these were held at Leiston Abbey or Belstead Brook, Ipswich. One weekend I completed a bricklaying course, on another I leant about tractor repair and maintenance, both of which proved useful back on the farm. The girls tended to be doing things like poultry trussing and flower arranging. However, I don't think it would have raised any concerns if, for example, one of them had wanted to do the bricklaying course, or a boy the flower arranging. In the evenings we had games, a chance for conversation, and dances, usually old time and square.

We held an annual public speaking competition, usually at Ipswich Town Hall. Teams of four competed in three different age groups from the various clubs. The teams comprised a chairman who introduced the subject, a speaker, and a proposer and a seconder of a vote of thanks. We spanned a variety of subjects, trying to inject a bit of humour into even the most serious topics, sometimes more successfully than others. I am not sure about Suffolk, but certainly on a national scale some of our agricultural leaders honed their debating skills at Young Farmers' public speaking competitions. One such was Henry Plumb, now Lord Plumb, who, through his involvement with the YFC, successively became president of the NFU, an MP, an MEP and president of the European Parliament before returning to the House of

Lords in addition to maintaining his farming interests in Warwickshire.

From time to time, we would arrange club outings. In 1951 we visited the Festival of Britain, not that I can remember much of the content now except for going to Battersea Park where there was a gigantic funfair. We sampled all the fun of the fair, which to us was the highlight of the visit, for it was still a time of austerity. On other occasions we would arrange a bus trip to a London theatre, arriving back home after the performance in the early hours.

We visited the world-renowned Agricultural Research Centre in Rothamsted, Hertfordshire, which was set up in 1843 by Sir John Lawes and Sir Henry Gilbert. One of the most famous experiments there was that on Broadbalk Field where winter wheat has been grown continuously since 1843. We had been brought up to believe that such a practice would lead to disaster, with a monoculture and no rotation allowing diseases to develop. The yield, where fertilisers had been applied and the crop hand-weeded, was quite respectable for the time. The original aim was to test the efficacy of organic and inorganic fertilisers on the yields, and the experiment continues to this day. Today, Broadbalk is still growing wheat, although it is now being treated to a much wider range of regimes, such as spraying for weeds and fungal diseases, ploughing in straw, removing straw, experimenting with different applications of fertiliser. This has contributed to an enormous database of information relating to wheat growing, which is available nowhere else in the world. Whether it can lead us on to better ways of combating weeds such as blackgrass remains to be seen.

As part of our wider interests, Stowmarket YFC visited Heathrow in 1953. With the rapid development of the aeroplane stimulated by the War, air travel was taking off. Heathrow was chosen as the site for a national airport, its proximity to London being a factor as well as there being room for expansion as the area then had a relatively low population density. It was a building site at the time of our visit and seems to have been one ever since, but as this was the future, we were keen to see it.

The social side of the YFC included tennis parties, cricket matches and treasure hunts in the summer. In the winter, the Stowmarket club hosted an annual dance, a black-tie 'do' held at the Old Con Club in Milton Road, Stowmarket. Even then, the Con Club had seen better days. However, by the time the boys had put up the decorations and the girls had swept up and arranged the flowers, we thought it looked quite glam. A popular local band, The Playmates, provided the music and Ken Woodward, one of our older members, was for many years our master of ceremonies. Ken enthused and encouraged us to take our partners for the Valletta, the St Bernard waltz or, more vigorously, for the Gay Gordons or the Dashing White Sergeant, as well as for a host of other dances. Usually, we finished at 1 o'clock in the morning, though one year we decided, as everything was going so well, to carry on for another hour. However, the general opinion afterwards was that it would have been better to stop while spirits were high than to have carried on and lose momentum. Another lesson learnt.

Stowmarket YFC also ran a Christmas party, a fun event with games as well as dancing to which members from other clubs were invited – Melton and Debenham, and, from over the county borders, Diss in Norfolk and Wix in Essex. We received reciprocal invitations which were enthusiastically taken up. We also arranged an annual county Young Farmers' Ball. One year it was at the Athenaeum in Bury St Edmunds, another at the Copdock Hotel near Ipswich. In 1956 there was a heating fault at Framlingham College which was due to be the venue. Lady Blanche Cobbold[1], who was at that time our County President, offered to host the Ball for us at her home, Glemham Hall. It was a very successful evening

---

[1] Lady Blanche was a daughter of the Duke of Devonshire and had married into the Cobbold family, a prominent Suffolk brewing family who were founder members and strong supporters of Ipswich Town Football Club.

reminiscent of the days when every stately home had its ballroom and when balls were part of the social scene. Another year we held the Ball at Lowestoft, after which some of us decided to go for a swim in the sea. The moon shone, the water was warm, luminescent and sparkled like liquid gold as we swam and splashed each other. It was great to be young.

The YFC was often known, not without reason, as the local marriage bureau. It was responsible for several romances and a not inconsiderable number of weddings. However, I met my wife Mary before she joined Stowmarket YFC at a party given by the Gooderham family in Bacton Village Hall, when I was 20. Mary was the daughter of a neighbouring farmer, William Hamilton of Cotton Hall. She was very attractive with dark hair and freckles. At the end of the Gooderhams' party, she stayed behind, as I did, to help clear up; we chatted, and I invited her out to the cinema.

Mary under Judas tree in the garden at Red House, 1955

We went to see *Shane* at the Gaumont in Ipswich. Mary remembers us having to queue for tickets as was usual in those days, but she began to feel unwell during the film. I drove her home and when I rang the next day to find out how she was, her mother told me that she was covered in spots and had chicken pox. She did not hold it against me, and the YFC figured largely in our developing romance.

Once a year, the various YFCs in the county would come together for the annual Rally. Originally it was an event to compete in showing off livestock which individual members had raised over the previous year. As time went on it widened to include livestock judging, cookery, flower arranging, dressmaking, poultry trussing, tug of war, tractor reversing and handling, even country dancing. It took place in different venues round the county. In 1952 I was involved in organising the Rally but unfortunately on the morning when we should have been bringing livestock onto the site at Chantry Park in Ipswich ready for judging, foot and mouth disease had been discovered in the eastern counties. As a result, all livestock movements were banned. I spent much of that morning on the telephone arranging for the YFC judging teams to visit the individual farms to do the judging rather than on the site at Chantry Park. The farmers who were due to provide the livestock were most supportive and fortunately the other activities of the Rally were able to go ahead on site and we were able to complete our schedule and successfully conclude the competitions.

Later that summer, because of the foot and mouth outbreak, all farm visits were stopped but we continued to meet weekly at non-farming venues, one of which was the local US Air Force base. Unfortunately, when 50 or so of us turned up, we found that our contact there had forgotten all about our appointment. We had to wait for an hour or so until he had been located and was able to assure the powers that be that we were not terrorists wishing to blockade or blow up the base. When we were eventually let in, we were hospitably entertained and found the visit interesting, not least seeing some of the aeroplanes which they were flying which seemed enormous to us at the time.

Through the YFC movement there were opportunities to participate in overseas exchange visits, some organised on an area basis. The first of these I joined was to Germany in 1951, the year before I met Mary. Nine or 10 of us, accompanied by Willy Coates, the YFC area organiser who arranged the trip, stayed with different families close to Hanover. We travelled by train and, although the War had been over for six years, the area round Osnabruck had been so badly bombed that it still looked utterly devastated. No doubt there had been more pressing priorities. We were assigned our various families. Fritz, my host, a lad of about my age, farmed with his family at Eldagsen. He used to get up at 5am, have his first breakfast, feed and exercise his horse and do chores around the farm, before his second breakfast at 8am. I joined him on a couple of mornings but after that I stayed in bed rather longer.

One day we visited the cement works in Springe where there were some very tall chimneys, each with a metal ladder all the way to the top. On spotting these chimney stacks Duncan Fearnley, one of our number, was reckless enough to say that he would give £5 to anyone who climbed to the top. I rose to the challenge. The next day I was very stiff but £5 richer, equivalent to my weekly wage at that time.

One Sunday some of us took the train to Hamelin where, walking beside the river Weser and wandering through the woods and the area surrounding Hamelin, we recalled the Pied Piper and how he had reputedly cleared the town not only of rats but also of children. Later we met up in a local hostelry with our German hosts who were drinking deeply of the local 'bier'. We remained there until 1 o'clock the next morning. They were very keen on their beer drinking but there were no concessions about getting up for the farm work at the usual time the following day. On their return visit I remember how pleased they were to visit the Greene King brewery in Bury St Edmunds. I tried to contact Fritz and his family about 10 years ago but was unable to find any leads.

In 1954 I had the chance to go to the US for six months as one of a party of 20 YFC members selected from around the UK. It was an International Farm Youth Exchange (IFYE) visit promoted by the US to foster international understanding and to enable Young Farmers from around the world to experience something of American rural life. They aimed to show us the activities of the 4-H youth organisation and the Extension Service, which was part of the US Department of Agriculture, as well as other youth programmes and local farm organisations. We were each to live with different farm families in various parts of the country[2]. I was assigned to spend my time in the states of Georgia and Michigan.

My diary entry for 4 June 1954, the day I embarked from Liverpool on the SS *Britannia* notes, '*There appear to be among the YF contingent 9 from England & Wales, 1 from Jersey, 3 from Scotland & 5 from Ireland.*'

It took 10 days to cross the Atlantic to New York, giving us the chance to get to know one another while on board and make friendships that have lasted to this day. I was not in a cabin with any of the other IFYE members but as I wrote to my parents from the ship,

I've a peculiar assortment of cabin mates, all Irishmen. One joined the ship at Liverpool. He is about 35 and doesn't seem to have stuck to any job in particular. During the War he was a cook aboard tankers and was torpedoed once in the channel. Since then, he has been storekeeper, commercial traveller, and a photographer on the promenade at Bangor... . Now he's off to the States with no particular job in mind.

---

[2] The US hosted delegates from around the world, including young farmers from England, Finland, Germany, Brazil, Uruguay, Sweden, Portugal, Greece and Israel. In exchange, that year approximately 125 US farm youth spent six months in 40 different countries.

Setting off from Liverpool for the USA, June 1954. I am in the back row, 4th from right, with John Cromie 3rd from the right, and John McGifford 6th from right

The other two came aboard the boat from Cobh, Eire on Saturday morning. I have not seen much of the one who seems to have a family aboard, but the other, an oldish man (70?), is quite a character but not very pleasant for all that. He is short with close cropped white hair & seems to be perpetually spitting. He hangs around the deck in a battered old felt hat & a dilapidated overcoat, occasionally going below to lay on his bunk dressed as he is. I don't believe he ever washes & has certainly never been out of his clothes apart from his socks, revealing a pair of dirty smelly feet. However, we're still alive & kicking.

I would meet up with others from the YF group throughout the day, on deck, for meals, for sing-songs and for dancing in the evenings. I got to know two of the YF lads from Northern Ireland well, John Cromie and John McGifford. Some 61 years later, when I had a hip replacement in 2015, John Cromie visited me in the West Suffolk Hospital, Bury St Edmunds while staying with his son who was working on a local farm. Meanwhile, as I was crossing the Atlantic very slowly westwards, my brother Robert was flying back from Guyana at the end of his National Service. I was sorry to have missed seeing him before I left. I just missed the call for National Service because when I was of age to be enlisted, farming was considered a reserved occupation. However, the rules changed a year or so later and Robert was called up. He joined the Royal Electrical and Mechanical Engineers, trained as a radio repair specialist and was sent with the Argyle and Sutherlands to Guyana where the Army was engaged in dealing with the rebel Cheddi Jagan. As with so many former colonies, the Guyanese were granted independence shortly after, and the first president was none other than Jagan. Meanwhile and less dramatically, Mary was studying in Eastbourne at the School of Domestic Economy. She and I corresponded regularly while I was abroad.

Much to the chagrin of some of the Americans on board, who had been telling us what a wonderful place it was, New York was fog bound on our arrival causing us to spend a day anchored close by the *Nantucket* lightship. However, I was up at 5am the next day to see the Statue of Liberty and catch my first glimpse of Manhattan before the ship docked safely at 7am. Disembarking 24 hours late, the IFYE contingent were whisked off by yellow taxi to Grand Central Station to board a train to Washington DC. It was the middle of June with temperatures up to 100°F and it was a new and strange experience to leave an air-conditioned building and hit a wall of hot air.

I spent a week in Washington attending a national 4-H club camp. The four Hs stand for head, heart, hands and health, signifying the whole body. As with everything in the States, the convention was on a grand scale. I learnt that to attend this camp is one of the greatest honours for any 4-H-er in the States. Only two boys and two girls from each of the 48 states were selected to take part, joining exchangees from all over the world. This included half of our group, who were detailed to stay, while the other half were promised a visit to another grand meeting in Kansas City at the beginning of July.

The week was spent studying the US government and its constitution, and being briefed on how we should comport ourselves while in the States. Some of it was rather long drawn out but there were visits to places of interest in and around Washington which were fascinating, and more than compensated for those induction sessions which dragged. We were given a tour of the White House and addressed by President Eisenhower himself. We visited the Capitol, the American counterpart to the Palace of Westminster, and one afternoon sailed down the Potomac river to Mount Vernon, the former residence of George Washington. In the evenings we had fun: went to the movies, the theatre and square dancing events. One evening we had a party at Washington Airport and had an opportunity to go on board planes which had stopped over en route from Texas to New York. On the last evening, a candle-lighting ceremony was held in the

President Eisenhower at the White House, 1954

The White House viewed from the top of
the Washington Monument, 1954

gardens of the Pan American building, a beautiful setting. Impressive it was as candles were lit and the 200 to 300 boys and girls filed around the large lily pond, their candle lights reflected in the water, much like the fireflies that flitted overhead.

Then it was all over but for the farewells. We were to disperse to our allocated states. In the morning I left for Georgia by Greyhound bus, travelling through Virginia and the Carolinas. I was interested to observe the changing landscape, from woods and forests, an occasional field of corn or tobacco, and haymaking and combines at work in Virginia, to the cotton fields of South Carolina. I arrived two days later in a sun-baked Georgia.

I spent one night in a hostel at the state's University of Athens, where I was lucky enough to be introduced to the person in charge of the pig section of animal husbandry on the university farm. He showed me their feed trials on Duroc pigs, which used irrigated and non-irrigated ladino clover, and experiments with different protein levels in the feed mix. I noted that some pigs were kept on slats in pens with no straw, something back home we were not yet doing, though as with so many things American, we eventually copied them. Through Grandad's friendship with Dr John Hammond from the University of Cambridge, I had been fortunate enough to be given letters of introduction to professors at several agricultural faculties at US universities. Athens was not one of them, but towards the end of my time in the US I was able to make use of two of the letters, enabling me to meet the professors of agriculture at the universities of Minnesota and Iowa.

The hospitality I received everywhere in the States was great. In both Georgia and Michigan I was received warmly into the families with which I stayed, treated as a family member, worked on their farms, played and shared their friends. I noted that we never went 'to' visit friends but instead would visit 'with' friends. The southern drawl, especially 'you all' contracted to 'y'awl', seemed to embody the region.

In addition to my time staying on various farms – four in Georgia and three in Michigan – I attended camps and a convention linked to the 4-H movement. Almost as soon as I arrived in Georgia I was enrolled in a naval stores camp near Valdosta, almost into Florida, naval stores being the name given to the gum that is obtained from certain pine trees and tapped in much the same way that rubber is tapped from the rubber tree. This sticky product was once used for caulking wooden sailing ships, hence its name, but by the 1950s the gum was mainly used in the production of turpentine. In a competition I won a hunting knife, which only recently I have surrendered as a dangerous weapon.

I soon became acquainted with southern fried chicken and the local patois. With temperatures reaching 110°F, air conditioning was eagerly sought. One sweated with the slightest exertion. Near Soperton in Treutlen County, I was taken out to fish for catfish from a small boat. Small wonder

Gathering tobacco, Georgia, USA 1954

Gathering tobacco, Georgia, 1954

Learning how to tap for gum at a 4-H camp, Georgia

Collecting gum (naval stores) in pan from pine trees

Naval stores gum being emptied into processing plant

in such heat that a favourite pastime was fishing in lakes shaded by pine trees festooned with Spanish moss. Another Georgian pastime (you could hardly call it an activity) was swinging or rocking on a chair in a screened veranda while visiting friends and drinking ice-cold tea. The grandmother of one of my host families would spend much of the day in a rocking chair on her veranda chewing tobacco and periodically spitting over the rail. This slow pace of life was so different from anything I had experienced before, though in contrast, the busy tobacco market in Vidalia was abuzz. Here, the never silent auctioneer led his buyers along the rows of baskets, his words running into each other so fast that one wondered how anyone knew what was going on. He averaged six lots a minute, but in one minute I saw him sell 10 lots. At the end of the row of baskets, he would take a slight pause, time for a cup of water, a wipe of the brow and he would be off again.

I was surprised to find that nearly half the state was covered in timber. I had read *Gone with the Wind* and was expecting to see large acreages of cotton, corn (maize) and tobacco. However, this view was mistaken. The first farm on which I stayed was 1,800 acres, 1,000 acres of which was forest, and the 800 arable acres were split into 80-acre lots and share-cropped[3]. A sharecropper and a mule or two to 80 acres was a common practice but the amount of tobacco and cotton they could grow was strictly limited. There was a serious drought in that 1954 summer and crops wilted and many sharecroppers lost money. While there, I took and passed a test for a permanent state driving licence. It entitled me to drive anywhere in the States enabling me to be rather more useful to my numerous hosts. The test was much less rigorous than at home. I filled in a form, answered some questions on the Georgian highway code, drove the patrolman round one block, all right-hand turns, paid one dollar and, as they say, Bob's your uncle.

I was fed on a variety of unfamiliar foods. I ate corn in all forms imaginable: corn bread, grits, corn on the cob, boiled corn and popcorn, beans of all shapes and sizes, and sweet potatoes. I never did get enthused about iced tea but that and Coca Cola were what one was generally offered. On the farms I was involved with mending machinery, milking cows, testing cows for TB, carting out muck, feeding broiler chickens (6,500 on one farm), laying a concrete floor in a barn, fencing by day and on several evenings was invited out to give talks to a variety of groups about farming in the UK. Having been advised in advance that this was likely, I had taken some slides of Red House Farm with me to show.

---

[3] Sharecropped or sharecropping: this was common in the southern US. A tenant farmer would work the land, being provided with accommodation, seed and tools, and receive an agreed share of the value of the crop less the cost of production. However, if the crop failed, the sharecropper still had to pay the costs.

I also had the opportunity to visit the state prison farm, a large-scale operation with such a big labour force that hardly any machinery was used. A silage clamp was being consolidated by a large gang of prisoners stamping on it. There I saw 'hogs' in quantity for the first time since my arrival in the States, the Montana breed being rather like the Large Black but with smaller heads. I was also taken on a tour of the prison itself, seeing the electric chair and its rather gruesome ancillaries – an embalming room and deep freezer drawers for the bodies.

Sometimes I was made the excuse for a special outing. It was on such a trip that I visited Warm Springs, Georgia where F.D. Roosevelt had stayed in the hope of easing, if not curing, his polio, and where he spent his last days. Nearby was the pleasure garden dedicated to Ida Cason. On another outing, I saw a drive-in cemetery for the first time, on the outskirts of Savannah, the oldest town in Georgia. I was reminded of the American serviceman who called one of our Suffolk towns 'half as big and twice as dead' as his local cemetery back in the States.

It was not until I moved to my third farm stay, Mountain Love Farm, Walker County, in the north-west of Georgia that I was able to work with pigs. I explained the set-up there in a letter home to my parents and brothers:

Sunday 22ⁿᵈ Aug. Mountain Love Farm, Walker Co. Ga.

Dear All,

At last, I have come across some pigs. On Thursday I moved from the Daffron's. Now, I am staying with the pigman Harry Heckerson, on one of the biggest farms in the state. The farm is owned by a soap manufacturer. There must be over 4,000 Acres, most of which is down to grass. Hereford cattle are the chief concern on the farm. Although there are only 5-600 head; they are looked after with one eye on the show ring. The same thing applies to the pigs. They are Durocs. To my mind they are ugly looking red animals. One boar here is nearly as high as he is long. However, they win prizes and can sell breeding stock at a little higher price. But it doesn't seem worth the extra trouble involved.

The pig farm is situated at the top of Lookout Mountain which is larger and flatter than one would think. There are about 30 sows and gilts and the progeny from them. About half the pigs born are kept for sale as breeding stock. I was due to have come here earlier in the week, but Harry has been showing pigs in Valdosta & was also in Miami from whence he sent about 20 pigs to Cuba.

Most of the pigs are kept outside with huts in the different lots if they want to go in. Feeding, except for those being especially prepared for shows, is by automatic dry feeders and there is a constant water supply to each lot. Apart from Harry there is one boy who helps with the hogs though he doesn't have much to do.

Interestingly, years later on our own farm, I was to introduce the Duroc pig, but all that will be covered in another chapter.

Farm hand feeding Duroc hogs at Mountain Love Farm, Georgia

One Sunday the Heckersons took me to visit Rock City. Built on top of Lookout Mountain, the city must have been one of the most extensively advertised in the States. For hundreds of miles, on every side, one saw signs and advertisements: 'See Seven States from Rock City', 'See Beautiful Rock City' and so on. With this great build-up, I had expected to see something more, or perhaps that is the art of advertising?

I found that religious observances in Georgia were more overt than at home. Seldom did one sit down to a meal without asking for a blessing. I stayed with both Baptists and members of the Church of Christ. With the latter, I often found myself arguing against their intolerance of other sects. They were enthusiastic Christians, as were so many of the people in both states I visited, with adults attending church, Bible study and Sunday School although some were strangely narrow-minded.

At the end of August, before leaving the South, I attended the state 4-H club convention, where I was able to renew old acquaintances and make new ones. There I was invited to contribute by saying a few words endorsing the IFYE scheme and its aims. It seemed to me that the Americans took great pleasure in constructing conventions and 'programs', though this one was, thankfully, shorter than the induction had been in Washington DC. Then it was time to move on. I was given a Greyhound bus ticket and travelled overnight to Michigan ready for a taste of farming in a northern state.

Michigan provided a contrast with the South, and apart from corn, the crops they grew were similar to those at home. There was mixed farming – cows, sheep, wheat and sugar beet – but also a large amount of timber and scrubland. I found many ways of helping on the three farms. When I arrived the wheat harvest was only just over and there was straw to be baled and carted and silos to be filled, and by November when I left there was still corn to be picked. In between I mucked out cow barns, spread muck, stone picked, made new corn cribs, extended the cow barn, cleaned eggs, ground meal, made new gates and so on.

Photo call for me as exchangee looking
at farm machinery, Georgia

Another photo call, Georgia

Cutting corn, Michigan

Mr Clemens unloading bales of wheat
straw on his farm in Michigan

I was fascinated to see part of the great Ford enterprise one weekend, visiting Greenfield Village and the Ford Museum. South Michigan was a great industrial centre in the 1950s. The US produced five million cars each year, of which 80 per cent were manufactured within one hundred miles of Detroit. The scale was mind-boggling.

Of all my experiences in the US the most memorable occurred during my last week. I had arranged to use the spare time at the end of my trip to stay with some relatives in Lansing, Michigan – Aunty Nancy and Uncle Clare Crandall, their married daughter Margaret and their two grandchildren[4]. One afternoon, Aunt Nancy and Margaret took me to visit a motor car plant in Flint, leaving Craig, the five-year-old, at home with his grandad. On the way home we stopped at Linden, the local town, to pick up the mail. A neighbour approached. 'Do you know that your barn is alight?' she said. No further words were wasted. We headed straight back to the farm. All possible ideas passed through the minds of my aunt and her daughter. What had happened? What was alight? Which barn was it? We could see the smoke and hastened on.

At last we arrived. Both barns were blazing heaps of ruins. The calves were burnt, and many chickens roasted alive. The fire engine, unable to do anything but prevent a nearby shed from catching fire too, had gone home. Uncle Clare and a few neighbours were helplessly gazing into the flames.

'Where's Craig?' Amid the tears the boy appeared from the house. The immense relief of his mother was apparent as she swept him up in her arms and embraced him. But how had the

---

[4] Aunty Nancy was the daughter of Alec Black, a first cousin to my grandfather, David. Alec Black had been farming at Westhorpe Hall, Suffolk in 1898 when Grandad first came to Red House Farm. Alec had a daughter, Nancy, and two sons, Bill and David, who were contemporaries of my father, Tom. From 1914 to 1916 they picked him up in their pony and trap on their way to Stowmarket where they all attended the secondary school. In 1916 Alec emigrated with his family to Michigan.

fire started? To Uncle Clare it was very confusing. He could only keep repeating, as more neighbours came on to the scene, how his grandson, running up to the house crying, had said there was a fire in the barn, how he had rushed over with a fire extinguisher and tried in vain to extinguish the flames. The rest we could see for ourselves. The suspicion was that Craig had been playing with matches. Before nightfall, several neighbours had offered help. One brought hay, for luckily the cows had been outside and so were safe; another brought feed for the pigs; another offered to take the chickens that had escaped. The next morning, they came again and converted one of the remaining sheds to hold the cows and even spoke about building another barn. It was encouraging to see so many rallying around and offering their assistance.

During my six months in the US, I encountered some misconceptions held by the Americans. Some thought of the British Isles as being very much larger than it is and some were surprised that we did not grow all the crops that they did in the States. Others thought that the English had no sense of humour but I was at least able to dispel that notion, and others suggested that in England we were very much controlled by the government, though it seemed to me that at that time the farmers in the US were more controlled than we, due largely to acreage restrictions on many crops, imposed because of over-production in previous years. Many of the people I met wanted to know how we were getting on with the then newly established National Health Service.

Possibly living in a country as large as the US tends to produce people who think big. Distance did not mean so much as in Britain. One day, while at a milk collecting centre, I took the offer of a ride into Detroit and back, a journey of about 70 miles each way. Perhaps by today's standards that is not far, but it was a length of trip that then I would only have made about half a dozen times a year from home, and rarely there and back in a day. But I did meet others who were strangely limited in their knowledge of anything outside their county let alone their state or the country.

Television was still something of a novelty for us in England but blared everywhere I went in the States. McCarthy was conducting his 'Reds under the Beds' witch hunt and the congressional hearings were constantly reported on the television news. McCarthy ruined the lives of many innocent men before enough folks came to realise he was only trying to embellish himself. He was eventually rumbled and was censured by the Senate in December 1954 just before I left for home.

I paid between 75c and $1.25 for a haircut while in the States, whereas back home it only cost me the equivalent of 15c, reflecting the relatively higher cost of labour in the US. As a result, staff were employed only when necessary. I noted a lot of the 'serve yourself' principle, whether in a cafeteria or an automatic machine into which one put a nickel, dime or quarter in exchange for a bottle of Coke, an ice cream or even a photograph. I saw one machine from which one could, after inserting the necessary change, obtain half a gallon of milk in a waxed container. Even the flat-rate charge for trams and underground journeys was an extension of the labour-saving principle, and I saw a parking meter for the first time. What was standard both on farms and in homes in the States, such as the refrigerator, the electric kettle and the television, took another 10 years or so to become widely used in England.

I returned home in December 1954, aboard the *Queen Mary*. This time I shared a cabin with the two Irish Young Farmers, John Cromie and John McGifford, whom I had got to know on the way out. We were able to exchange experiences. It was a rough crossing, with five particularly stormy days. I would not have believed that a storm could be strong enough to stave in the portholes of the first-class dining saloon; fortunately, all we had to contend with in third class was the crockery sliding from one side of the table to the other and the periodic spillage of drinks. The weather was not much kinder when we landed at Southampton though at least we had some inkling of what we were likely to find on our

Self-service, new to me in 1954

Impressive scale of the stockyards in Omaha

Memorial stone to the Black family who emigrated to
Michigan from Westhorpe Hall in 1916

SS Queen Mary in New York harbour. I returned home
on board in December 1954

return. I felt sympathy for the many Jamaicans who had travelled with us, hoping for a better life in England. They seemed more dressed for the Caribbean than for the winter chill and grey skies that greeted us all, or for the 'pea-souper' they would encounter when they reached London.

Back home, I continued to participate in Young Farmers' events and in 1958 was elected YFC county chairman. It was an interesting year. The county organisations were coming under increasing financial pressures and I became closely involved in the discussions which led to the merging of the East and West Suffolk Federations of YFCs. I was elected first chairman of the combined Suffolk Federation and served for a year or two before relinquishing the role as my other responsibilities grew.

Me, back home, attending a YFC meeting, 1955

On a visit with Stowmarket YFC to the Houses of
Parliament in 1955, hosted by our local MP, Harwood Harrison.
(I am in back row, 3rd from left)

CHAPTER 5

BROADENING HORIZONS

Denmark had been a big supplier of bacon to the English market before the War, and its pig industry was now making a strong recovery. English farmers wanted to know what it was that put the Danes in such a strong position in our market. In 1949 Father and Mother went to Denmark with a group of local farmers. Father was impressed with what he saw while there and, on his return, suggested that I too should go to Denmark to learn what I could. Maybe we could find ways of recovering more of the market for ourselves.

So it was that I and Jim Henderson, deputy manager at the Elmswell bacon factory, joined a party of farmers from Kent on a trip to Denmark in 1951. We were both keen to see what was going on there and to learn what we could. On visits to several pig units, we were impressed with the standard of livestock care, with usually the whole family, husband, wife, and children, being involved on the farms.

The Danes were very proud of their breed of pig, the Danish Landrace. It was a longer pig than most of the English breeds and had less back fat. During the War, British customers were pleased to get whatever meat they could; the amount of fat on it was of little significance. In the early 1950s meat was still in short supply and it did not come off ration until 1954, but there was now a move against eating too much fat. Consequently, there was market pressure to reduce the back fat on our English pigs. The Danish pig was better suited for this market, and it was also reputed to be a better converter of food into meat.

Before the War, the Danes had instituted a national progeny testing scheme to improve the genetics for the quality and performance of their pigs, building five testing stations around the country. The project was overseen by a Hjalmar Clausen, who reinvigorated it after the War. We were privileged to have him speak to our group. He believed strongly that progeny testing was the best way to improve the genetics of the pig and presented statistics to show how their programme of pure breeding, with progeny testing, was the right way forward. He was adamant, however, that they would not be prepared to export any pigs lest they lose the competitive advantage which they had spent so much time and effort building up.

One day we took the ferry to Scania, the southern part of Sweden just over the water from Copenhagen, to visit pig operations there. These were much like those we had seen in Denmark and, although the genetics were not identical to those in Denmark, they were similar, and we learnt that they would be prepared to export pigs to the UK, more of which later. Two crops that I observed with interest in Sweden were oil seed rape and vining (shelling) peas. Birds Eye was already freezing peas in the UK, but rape was not to make an impact back home for another 30 years.

One evening we were joined at dinner by my cousin Brian and his wife Karen. In the Royal Navy during World War II, Brian had been an engineer on the cruiser *Diadem*, escorting convoys to the Russian port of Murmansk. VE day found him celebrating with his shipmates in Copenhagen. There he had met Karen, who took him to her home to meet her parents. They got on well and he decided he would settle there, learn the language and take an engineering degree at the University of Copenhagen. They married and he began working for Atlas, a Danish machinery company specialising in refrigeration worldwide, later moving on to DFDS, the Danish steamship company. He worked his way up, eventually becoming general manager with worldwide responsibilities. Wherever we went as Brits in Denmark we were received with a rapturous welcome. The War may have been over for six years, but we were considered saviours of the country and treated as such. It made me feel very proud and yet very humble. It was my first visit to that country which I have returned to many times since.

It must have been shortly before that trip to Denmark that I first met Stephen Horvat who was to play a large part in my life over the next 20 years. He had unbounded enthusiasm and taught me a great deal. Stephen was born in 1911 in Czechoslovakia into a landed family. He was at the University in Berlin at the time of the rise of Hitler. Impressed and appalled, he saw Kristallnacht at first hand. His father sent him to England, maybe with concerns as to what was to come, for it was while Stephen was here that Hitler invaded Czechoslovakia. Stephen was advised not to return home although he narrowly avoided being sent back when his visa expired. However, following the invasion of Poland in September 1939 and England declaring war on Germany, it was out of the question that he return to his homeland. Stephen joined the Army and served in the tank corps throughout the War. He was always something of a charmer and married an English girl, Rosalind.

After the War, Stephen returned home to the family estate in Czechoslovakia with Rosalind and their son Mark. The farm was still there, but Stephen's parents had disappeared. What had happened to them in the intervening years, nobody knew; one day they were there, the next they were gone, doubtless victims of the gestapo. The farm was large and Stephen, having reclaimed it after the War, employed both a gardener and a chauffeur. Unfortunately they were both ardent members of the Communist Party, actively pursuing the communist dream that was spreading all over Eastern Europe. Landowners particularly were in their sights as 'enemies of the people', and at the behest and encouragement of the Party, Stephen's farm was all but taken over by these two. Fearing for the future, Stephen sent Rosalind and Mark back to England in 1948 or 1949 in a private plane piloted by a relative. Matters grew worse and having secretly organised to have various possessions shipped out of the country, Stephen secured a permit to leave Czechoslovakia to visit the Royal Show in England and left with just a briefcase. He said that the walk across the tarmac to board the plane was the longest walk of his life, fearing a hand on his shoulder if the authorities had discovered his plan never to return. Many years later he said to me, 'You can't frighten me, David; I've been to the edge, looked over and survived.'

When I first met Stephen, he was keeping pigs and poultry on a small farm at Worlingworth, chopping fodder beet to help reduce the cost of the ration. He was a man bubbling over with ideas and usually had four or five projects going at the same time. Later, he was to say to me that he would have ten ideas a day and if one of those ideas bore fruit, he would be happy. More than most farmers at that time he could see how times were changing and he was determined that he was going to be part of that change. Landrace pigs, Framlingham Farmers (Fram Farmers), Elmswell bacon factory, the Farm Buildings Association, the Hysterectomy Pig Company were all areas into which Stephen threw himself with infectious vigour and visionary enthusiasm. Stephen told me that his father used to spend three or four weeks each year travelling round Europe visiting farms and finding out more about the latest developments in agriculture and politics. He reckoned one could make more money off the farm than on it, something I have often pondered having been brought up to keep my nose close to the grindstone.

Pigs were a large part of our business at Bacton and we were looking to improve our stock by using bloodlines from Scandinavia. In 1949 the Government had sanctioned the importation of three batches of Landrace gilts and boars from Sweden. Stephen was anxious to see how they had made out in British conditions, for we felt they had much to offer the UK pig industry. One batch was on the Bluegate Farm, a Strutt & Parker farm in Essex, another was in Yorkshire and the third in Dyce in Scotland.

It was arranged that Stephen and I would go on a fact-finding expedition to see these pigs in Yorkshire and Scotland. Our trip was most successful. Stephen had an almost inexhaustible supply of stories and anecdotes and the journey whizzed by. At the Yorkshire farm we were warmly welcomed and shown the Landrace pigs by Henry Wertheimer and his pigman. At one point, however, Stephen was sternly admonished for asking questions of the pigman. 'You vill not ask my pigman, you vill talk to me', said Henry. I am not sure that he had anything to hide but it did sound a little strange. Stephen excused himself to me and then spoke back forcibly in German no doubt thinking Henry would understand him better in that language than in English. The Swedish Landrace pigs were doing well on the farm and it appeared that the finishers that had been sent to the factory were living up to the expectations of producing better bacon by being longer and having less back fat than the farm's other pigs.

On we drove to Edinburgh to see the Miss Parks, a delightful pair of sisters who ran a well-respected herd of prize-winning pedigree Large Whites. We were shown round an immaculately presented operation and then entertained to a traditional Scottish afternoon tea while discussing the future of the pig industry.

That evening, having time on our hands, we went to the cinema. It was showing *House of Wax*. It was one of those blood-curdling horror films. We had to don a pair of glasses with one lens red and the other green, so that the events could be seen in 3-D. That was a new experience for us both. It led next day to Stephen regaling me, on our way to Aberdeen, with tales of Vlad the Impaler and Bluebeard, stories that he had been told as a child.

We continued to Dyce, just north of Aberdeen where our hosts were the Lawson brothers who were not only pig keepers but also ran the local bacon factory. There, they showed us carcasses hanging on the hook, giving us a chance to compare pure bred Landrace pigs and cross-bred Landrace x Large White against their usual run of various breeds and crosses. We could see the benefits of the use of these pure and cross-bred Landrace pigs, which were producing better quality bacon. They were longer, had less back fat and had a larger eye muscle, essential for good bacon. The Lawsons were enthusiastic about the newly imported Landrace pigs but felt there was a need for a greater number to be brought in to prevent inbreeding. They were agitating to persuade the Scottish and English ministries of agriculture to grant licences for more to be imported and asked that we add our voices too. The campaign was successful and in 1954 another 100 Swedish pigs were brought in and sold at auction, but more of that later.

We had another warm welcome from Alex Alexander, a Large White breeder who delighted in showing us round his pig operation before we drove on to Gordonstoun, where Stephen was considering sending his son Mark to continue his schooling and education. It was a new experience for me to be sitting on the other side interviewing a headmaster rather than the other way about. However, we got along well and Stephen was obviously sufficiently impressed to let his son Mark spend many of his formative years in that rather spartan environment. Back home from the trip, I could understand better what Boswell must have felt on returning from another Scottish journey with his mentor and friend Doctor Johnson.

After World War II ended in 1945, it was generally agreed that British agriculture should provide a greater proportion of the nation's food requirements. To do so, it needed some sizeable investment: little had been spent on repairs to and renewals of buildings and machinery for six years. New ideas in areas such as genetics, nutrition and labour efficiencies needed developing and putting into practice. Farming was supported and encouraged in a way never seen before.

With grant aid from the government, there was an explosion in research and development. Modern buildings were needed for a wide range of uses including livestock production, grain storage, workshops and so on, but to what designs should they be built? It was to help answer these questions that the Farm Buildings Association (FBA) came into existence in 1956. The initiative came from a group centred at the Rowett Institute which was part of the University of Aberdeen. The Mackies, a far-sighted farming family from the Aberdeen area, and David Soutar from the university, were prime movers. They brought together farmers, building manufacturers, architects, veterinarians, structural engineers and researchers in all these fields, and civil servants and politicians. It was a mix that stimulated ideas and practices which had been on hold over the War years.

I was keen that we join the FBA and we became early members of the organisation, as did Stephen Horvat. Through the FBA I travelled extensively looking at new developments in pig buildings at home and overseas: Denmark, Ireland, Italy, Scotland, Germany, the US and Sweden, to see some of the new designs in practice. Grandad, on my return, would always ask, 'Well, what have you learnt?' and it was considered terribly bad if one was unable to list two or three ideas to benefit the business. Fortunately I usually came home with some useful ideas or concepts which I had either seen or picked up through conversations with others. Developing and putting into practice new ideas made for exciting times.

On a trip with the FBA to Northern Ireland in 1958 we called in to see the McGuckian brothers, leading pig keepers and businessmen from Ballymena. At that time, they were able to obtain penicillin, which was not yet freely available in England. They were using it very effectively to treat pneumonia and other pig diseases and, when added to pig diets, it had the effect of improving conversion rates. We brought some home with us and we too found it very effective. Shortly after that, it became available in Suffolk and we used both penicillin and later another antibiotic, aureomycin, extensively. The use of antibiotics generally has subsequently become a very controversial issue, but they served us well. No doubt that when something works well there is a tendency to overdo its usage and it is only later that one becomes aware that there might also be negative effects. At David Black & Son we still use antibiotics, but only sparingly and our usage is now well below the advisory limits, made possible by careful management, hygiene and an all-in, all-out system[1] which we have gradually introduced.

It was through Stephen that I came to know Bill Green. He and his brothers were not born of farming stock but were determined to be in agriculture. Bill's two brothers saw an opportunity in producing salads and vegetables for the supermarkets and built a big business based in Soham, Cambridgeshire. Bill felt that chickens offered great business opportunities and had established a chicken operation at Badingham. Together with four or five other local chicken producers he had organised the finances and built a processing plant at Eye. Pollastra, for so it was called, was in many ways a similar operation to that which we had for pigs at Elmswell, a co-operative, handling members' produce and developing links with the rapidly expanding supermarkets. The pressures and stresses of running a developing production and processing business supplying the expanding and demanding supermarkets, as well as accessing the funds needed to do so, meant that we had many interesting discussions together.

Bill was a founder member of a discussion group of 12 or so like-minded farming enthusiasts known as the 75 Club and in 1976 he invited me to join. I have now been a member for nearly 50 years and have been chairman on a couple of occasions. It has been a great experience hearing from our members as well as our many varied speakers. These have

---

[1] All-in, all-out system: where each pig-fattening shed is emptied of pigs approximately every fourteen weeks and thoroughly cleaned before the next batch of pigs is brought in to be fattened.

included members of the royal family, prime ministers, agricultural ministers, governors of the Bank of England, chairmen of British businesses, overseas ambassadors, and a wide range of others, often contacts of one or more of the members. Not surprisingly over the years there have been many changes in the membership. We have had Dutch, German and French members and profited from visits made to their countries; membership has expanded and not many of the original group are still alive, but it still fulfils the original concept of 'dealing with mutual problems affecting us'.

Of the speakers at the 75 Club, I can only give a flavour here. I found Edward Heath particularly difficult to engage with. Most people, when one speaks with them, at least make encouraging noises or gestures even if they say nothing. He listened in complete silence and waited until I had finished my remarks before making any response at all. As a country boy I found this rather off putting. In complete contrast, Princess Anne who joined us at one of the 75 Club's 'parent and offspring' evenings held at the Farmers Club, which my daughter Christine attended with me, was a much more comfortable and personable conversationalist. This was despite, or maybe because of, the many people to whom she must have been introduced in her lifetime. At the end of the evening, we presented her with a tree which, we hope, still flowers in Gatcombe Park.

I had been at school with David Plastow and invited him to speak at one of our meetings. David had begun his working life as an apprentice at Vauxhall motors but by the age of 40 had become managing director of Rolls Royce. He was responsible for the reverse takeover of Vickers, the engineering conglomerate. In his talk to us at the 75 Club he was able to shed some light on the ramifications of big businesses and their interactions with government. Another captain of industry we invited was Sir John Harvey Jones from ICI. He had fronted the *Troubleshooter* series on BBC television, which advised companies on how they should develop their businesses. One of these was Copella, an apple juice company from Boxford, south Suffolk, founded and run by an entrepreneurial lady, Mrs Devora Peake and her family. She had been born near Tel Aviv to Russian immigrants.

They had been growing apples in Suffolk since 1938. Mrs Peake was a driving force and I invited her to this meeting as my guest. We were entertained to a ding dong between her and John Harvey Jones, especially of note since she subsequently built a very successful business by ignoring some of his advice. Clive Jenkins, then the general secretary of the Association of Scientific, Technical and Managerial Staffs, a trade union, was another memorable speaker. He enquired if we had any press present. When we told him that everything was conducted under Chatham House rules, he wondered, half humorously, why he had come. It was Sir Hector Laing, chairman of United Biscuits who, in discussing Britain's future, said that we would be likely to hear a lot more about a certain lady called Margaret Thatcher.

With the 75 Club, Mary and I have visited many European countries where we have had the opportunity to see many farms and enterprises, to meet fascinating people and to enjoy the cuisine of the specific region we have been visiting. It has been a very rewarding experience. One of the current members, Hubert de Vrindt, farms to the north-east of Rheims in France, on the Chemin des Dames. Like us in East Anglia, he runs a mixed arable and livestock enterprise on a similar scale. The Chemin des Dames is a strategic ridge some 30km long, which was bitterly fought over and under, as a local museum attests, in both world wars. Relics are still being found. Close to the Champagne region, he has organised several enjoyable visits to farms and vineyards through his contacts in the area.

Fast forward to 2000 and it was through conversations with a fellow member of the 75 Club, Simon Scott, that I felt confident in pursuing the purchase of a different enterprise for the business which we named Portable Space. I shall say a little more about that when I mention the next generation.

Throughout my farming life, visits to other farms both in the UK and overseas often helped inform decisions and shape my views and we were then able to incorporate ideas as appropriate on our own farm. I have always been impressed by how farmers have been prepared to give of their time and discuss their businesses' successes and failures with other farmers, and long may that continue. When there was something specific that I wished to explore I would often plan a trip, arranging to call on several farms. In 1979 for example we were considering building flat decks and introducing an 'at a glance' pig recording system. Knowing of farms in the Holderness region of Yorkshire using these, Nigel, our pig manager, and I took off to visit them and, encouraged by how well both the flat decks and the recording system operated, we felt able to press on with our plans for them.

Overseas travel has always had a fascination for both Mary and me. Many of the trips have been with farming groups which have given us the chance to see a country and its people at a more intimate level than the usual run-of-the-mill travel package. These groups held another attraction, namely that of spending time with fellow UK farmers which often led to relationships which have lasted over the generations. We hope, as time goes by, that we shall continue to be fit enough to carry on enjoying the sights and the company as we have in the past. Talking to the locals helps one understand what makes things tick. In the farming world it is surprising how many concerns are universal, whether it is coping with the weather, bureaucracy, finding one's markets, controlling weeds or diseases, and so on. In China in 1979, even though we did not understand the language, we often clicked what our host was trying to explain faster than our interpreter who was not a farmer. Of course, it was not just farming that we saw on these holidays, for we have also been fortunate enough to have visited many of the world's great sights, such as the Terracotta Army, the Taj Mahal, Victoria Falls, the Temples of the Nile.

In 1975, while Mary and I were in the States, we spent some time in Champaign Urbana, Illinois with Professor Arthur Muehling of the Department of Agricultural Engineering and his wife Mary. He introduced me to local pig farmers whose farms I could then visit. Several years later I wrote to Professor Muehling with a few questions, among which I was keen to know whether the US had taken up the 'teardrop' system for slurry handling and, if so, to what degree. It was a system I had seen in use in Australia in February 1983, but it had only just been completed. The teardrops were slots in the concrete floor running the length of the building, which was on a slope. The idea was that the slurry would run into these teardrops and could then be washed out of the shed more easily. I thought it had possibilities for our set-up. He wrote back, with little enthusiasm for it, saying, *'I'm sorry, but I can't encourage you much about the teardrop system. If it were a cheaper and a better way to go, we would see more use of it in the US.'* Needless to say, we did not pursue this scheme, but this is just one example of when it proved helpful to have contacts outside the UK with whom I could sound out ideas.

In 1990 I was invited to a meeting in Denmark of European pig producers and suppliers to the industry to discuss the present and the prospects for the pig industry in Europe in the future. It was so successful that a group of us decided it would be worthwhile meeting together on an annual basis. The idea caught on and thus the European Pig Producers (EPP) association came into being. We were fortunate in our first president, Philip von dem Busshe, who was bilingual and president of the Deutsche Landwirtschafts Gesellschafte, the big German agricultural organisation, which he persuaded to support this new association by providing secretarial assistance, which it has done ever since.

The annual meeting of the EPP takes place in different venues around Europe. Some three or four hundred pig producers and suppliers of pig equipment come together for three days to exchange views, hear about the latest developments and learn something of the pig business in the host country. Most of the business is conducted in English or German which may not be ideal for everyone, but it seems to work. We have a large membership of Danes, Dutch and Germans, as well as a few members from the other European countries and even from South Africa. Mary and I have attended most of these meetings and made some good friends. Each year it has been held in a different country, each interesting in its own way, and has given us an introduction to places that otherwise we might never have visited. The Covid-19 pandemic, however, meant there was no EPP meeting in 2020 nor in 2021.

In their own way, each of these opportunities and experiences made me much more aware of the world beyond Bacton and, despite superficial differences, of how much we have in common with folks the world over.

Meeting the NFU President in 1967. From left: DB, Peter Shearer, Don Alston, Sir Gwilym Williams

Presentation to Stephen Horvat by Antony Daniell, 1970 (DB half hidden far right; Rosalind Horvat, far left)

Mary and I at European Pig Producers Congress, Bruges, 2015

CHAPTER 6

PIGS

My father was the one who, in the 1920s, started keeping pigs at Red House Farm. As a young man he did not enjoy having to milk the cows twice a day, seven days a week. He called on a neighbouring farmer, Fred Finbow at Coppings Corner one evening and was shown two sows, each with litters of 10 pigs and learnt that a sow could produce two litters each year. The pigs could then be sold on immediately or kept and fattened, before being sold after they had put on weight. The demand was there, for there were several dealers who travelled around buying pigs directly from farms. Alternatively, they could be sent to one of the local markets or a deal could be done directly with a local butcher. 'There's a profit in pigs, if you can keep them alive', he was told.

Sufficiently impressed, Father prevailed on his father to buy some 30 weaners and see what he could do with fattening them. 'Alright', said Grandad 'but you must keep a record of how much you spend on them.' Four months later the pigs were sold to Mr Hart, a Stowmarket butcher, and Father was able to show that he had made a profit. It was sufficient for him to repeat the experiment and, in addition to buying more weaners, he invested in two sows. So began our pig business.

Initially the pigs were kept in the old cattle yards bedded on straw, with feeding troughs made from cut-down forty-gallon barrels, but in 1926 a shed especially for pigs was built. The pigs must have done well because by the late 1920s all the cows were sold, and the monthly milk cheque was replaced with a weekly pig cheque. From time to time, Father would sell his pigs in the market, but he never felt very happy doing so; there was talk of buying rings and sharp practice, so he soon started selling them to the co-operatively owned bacon factory at Elmswell, more comfortable for knowing exactly what they would get for each pig sold. That was the beginning of a relationship with the factory that lasted until its closure some 70 years later.

Some people made a living dealing in pigs and would go round the local farms buying and selling pigs of all sizes. In common with other pig dealers, Harry Clarke who lived in Old Newton, always had a good story to tell and boasted that he was the biggest pig dealer in Suffolk, with some justification as, apparently, he weighed over 20 stone. A lot of tales surrounded Harry. He remarked of a fellow dealer, who was boasting of having made a particularly good deal, that he couldn't keep his mouth shut even if he was under water. On another occasion, Harry was heard saying to his friends in the pub after a good market day, 'I want you all to have a drink on me at my funeral.'

'Well,' said one of them, 'Harry, shall we be having it on the way to the funeral or on the way back.'

'You duzzy ol' fool,' replied Harry. 'That'll have to be on the way there. I shan't be with you on the way back.'

There was much to learn but there was little information about the best ways of feeding, housing and breeding pigs. Pigs were viewed as scavengers. Where they were kept on farms they were often out on grass, being fed household waste including potato peelings, or fed the same as the cows: mangolds or sugar beet tops with maybe a little grain and beans. Fortunately, Grandad had many contacts in the farming world, one of whom was a Mr Todd who had been experimenting with different feed rations. He had found that a ration consisting of 65 per cent barley meal, 25 per cent middlings and 10 per cent fish meal suited his pigs better than any of the other mixes he had tried. The barley provided the carbohydrate, the middlings the roughage and the fish meal the protein and minerals. Accordingly, Blacks tried it too. It worked so well that, although minor changes were made over the ensuing years, it formed the base on which our feed was formulated for a long time.

An ex-Army hut was purchased in which to mill and mix feed for this expanding pig enterprise. It was erected on the Red House site and in it was installed a mill to grind both barley for the pigs and oats for the horses. Fish meal and middlings came in sacks. The rations were mixed on the floor by hand and then were shovelled into 12 stone bags. The meal would be tipped into a tub, mixed with water and fed as a gruel. There were discussions about whether it was best mixed and left to stand overnight or fed freshly mixed. If it was mixed overnight, unless it was properly stirred in the morning, the first pigs to be fed had the lighter ingredients and more water, and the last pigs did that much better. A bucket, or maybe two, was used to tip the meal from the tub into a round trough, probably an iron one cast locally, in the middle of the pen. It was not surprising that, from time to time, a pig man would be bowled over by a bunch of hungry pigs anxious for their breakfast. Other food such as mangolds or beet tops would be thrown over the pen wall.

A lot of hand work was involved in producing the feed so, in 1934, a new building we called the grinding shed was erected to make it easier. Barley arrived in sacks by lorry, an innovation because, until then, the haulage had been done by horse and wagon. The sacks were hoisted up on a chain from the lorry onto the first floor of this new shed and the grain tipped into a hopper above the grinding mill, the ground corn then being blown into a holding bin. The other ingredients were conveniently stacked around a hopper into which the appropriate amount for a one-ton mix was tipped, conveyed up an elevator into the mixer and then into bags on the first floor ready for dispatch to the pig units.

This new, bigger mill in the grinding shed was driven by a 45hp Peter Brotherhood diesel engine which was started using compressed air. Occasionally the air would run out before the engine roared into life. Then four or five men would be called on to straddle the belt connecting the engine to the countershaft and pull on it until the engine started, not something that would be approved of today by the health and safety lobby. The countershaft also drove a compressor to recharge the compressed air cylinder and an electric generator. This building and set-up coped well with the increasing demand for pig feed up to World War II.

Before the War, Father had built farrowing[1] sheds catering for about 100 sows at each of the Wickham, Norton and Mendlesham farms, and fattening sheds at Bacton. These replaced several odd pens in a variety of old buildings. The bringing together of larger numbers of pigs created other problems, or perhaps it was just that it enabled them to be more easily seen. Pneumonia, erysipelas[2], roundworms and swine fever were some of the common ailments. Little was known about the causes of these diseases but there were treatments, some more effective than others: M & B tablets for pneumonia, a serum against erysipelas, and roundworms were treated by pouring oil of chenapodium down the pig's throat. Both swine fever and foot and mouth disease were notifiable which meant that, if either of these was diagnosed, the Ministry of Agriculture had to be informed. This resulted in all movements from that farm and those in the near vicinity being stopped. Infected and contact pigs were slaughtered on the spot. In 1938 our pig unit at Abbey Farm in Wickham Skeith was struck with swine fever resulting in some 100 sows being slaughtered. With the onset of the War in 1939 the use of grain for animal feed was severely restricted; it was deemed more appropriate to feed the cereals to people rather than to livestock. The effect of this on our pig numbers is evident: in 1939 we sold a total of 3,339 pigs but by 1945 the number of pigs sold in the year was only 317. We had a small allocation of cereal for the pigs

---

[1] Farrowing: giving birth.
[2] Erysipelas: a bacterial infection of the skin, which can cause problems in the lymph system if it progresses unchecked.

which we supplemented with home-grown fodder beet and chat potatoes, the small, discarded potatoes from the grading and sorting process. After the War, there was a government scheme to ensure there were sufficient potatoes to feed the population even in a lean year. Surplus potatoes from a good year were bought up and dyed purple marking them as animal feed and to ensure they did not re-enter the human food chain. The supply was erratic, but for several years we were able to buy these potatoes and steam them in one-ton bins. The pigs did well on them, and a hot potato was appreciated as an addition to lunch by those members of staff working close by.

It was a great step forward in 1946 when mains electricity came to the farm. True, we had to replace the old DC motors in the grinding shed and the workshop with new AC ones, but it meant we no longer had the problems associated with carbon brushes and commutators that went with them, nor the maintenance of the generators. We were

Steaming potatoes for pig feed, 1943

Herbie Betts and Albert Barker weighing pigs, 1934

Pigman, John Bloomfield, feeding pigs, 1956

no longer restricted by the amount of electricity we could use, nor the size of the motors and, needing to up the output of feed as we increased pig numbers, we were able to replace the diesel engine which powered the mill with a higher-horsepower electric motor. Our pig records show that by 1953 we had built the herd up again, selling 3,653 pigs in that year.

Although meal was still carted out in bags, we were gradually changing over to bulk handling. To complete that change, we needed bins and silos for storing grain and meal, and in 1950 extended the grinding shed. There was a large concrete base where the old diesel engine had stood. We needed it removed so there would be space to extend. We brought in Stan Fellingham, a local contractor who had a great dealt of experience in blowing up tree stumps. Stan drilled holes in the concrete and filled them with gelignite. Robert and I took cover with Stan as he lit a long fuse. It went up with an enormous bang, bits of concrete were thrown all over the place even perforating the old, corrugated iron roof but, at least we could now handle the job with picks and shovels and get on with the extension.

The shift to bulk handling also meant that we needed trailers capable of discharging meal into new bulk bins on the various pig units. We had these trailers built to a specific size by Rowlands of Aylsham, as they needed to be low enough to go under the railway bridge on the road through Bacton. Over the years we extended the grain storage facilities, installing bulk bins to hold the various feed-mix ingredients over a weigh hopper, giving us greater flexibility and faster mixing. Ingredients for the feed mix included tapioca, fish meal, soya and ground grain.

Another step forward was when we installed a 40-ton weighbridge in 1968 which helped us in several ways. We could measure more accurately the yields of our crops on the individual fields, we could weigh exactly how much meal went to each of the pig units thus removing some of the guesswork in conversion rates, and with more products coming onto and going off the farm in bulk, it made sense to be able to check that incoming loads were as detailed on the delivery note. Occasionally we had suspicions that this was not always so.

My visit to the States in 1954 had shown me that there was potential for livestock development and I saw large-scale operations in cattle, pigs and poultry; existing practices were being questioned. The gestation period for pigs is three months, three weeks, three days, and allowing eight weeks suckling on the sow, it was possible to have two litters per year. At Iowa State University they were experimenting with weaning piglets at three days of age instead of the eight weeks. The incentive for this early weaning was to bring the sows back into pig sooner and so produce more piglets per sow per year. The new-born pigs were given a chance to suckle and get colostrum with its antibodies into their systems before they were taken off the sow and fed on a largely dried milk diet before changing over to the more normal creep food. On returning home from the US, I found that BOCM (a British animal feed company) had taken up this idea and was producing a specific milk replacer, Sowlac. I persuaded our pigmen to give it a try, but we had problems getting the sows back into pig so soon after farrowing. We were not alone, and this practice of very early weaning died out. However, this kind of experimentation using improved nutrition for both sows and piglets led to the weaning age dropping to three or four weeks which is now the industry norm. Other universities were experimenting with genetics both theoretically and practically, amassing large quantities of data which was being processed on mechanical calculators. Progress was slow then, but the arrival of the computer, with its ability to handle ever more information, has enabled the whole science of genetics to develop much more rapidly.

We had already adapted several of the old buildings on the farm for pig housing, including those at New Barn Farm which we had bought a few years earlier from Joe Dennis. Joe had retired and had no family to carry it on. It was 60 acres and joined onto the farmland at Red House. Two of the buildings were cheaply converted to take finishing pigs bedded on deep straw, allowing us to scrap two older, ramshackle buildings which we always referred to as 'the Jubilee' and 'the Coronation' having been put up respectively in 1935 on the Silver Jubilee of King George V and in 1937, George VI's coronation year. Before we started work at New Barn we had to sort out the water supply as Joe's only source of water for both drinking and washing was stored rainwater. It was in the days before we needed a licence to sink a bore hole. Thurstons of Old Newton had already sunk wells for us elsewhere on the farm and so we brought them in to drill another one here.

In 1955 a new farrowing shed was built at Father's instigation at College Farm, Earls Green, which we had purchased the previous year. This was to cater for 120 sows with the pregnant mothers running in meadows outside. However, after a heavy rain, the only entrance to the sow yard became a quagmire. We struggled on with the system, shutting the sows in during the wettest of weathers, but it did make us query the benefit of allowing them out at all. This shed was built very much on the lines of our pre-War farrowing sheds which had no insulation in the roofing. It needed a lot of straw to keep the pigs warm and muck still had to be cleaned out by hand every other day into a trailer pulled by one of those little grey Ferguson tractors. In each pen we incorporated troughs along the central passage to simplify feeding, and nipple drinkers giving the sows drinking water on demand. Farrowing rails and a creep feed area with infrared heater were fitted. Infrared heaters were relatively new to us, as mains electricity had only recently been laid on to the units in which we housed our breeding herds. The warmth encouraged the piglets to move to the creep area after suckling, reducing the risk of them being lain on by the sow.

It was in this farrowing shed at College Farm that we first were stricken with a nasty disease known as transmittable gastroenteritis. Why it should have appeared here we never found out. On 23 January 1964 first one pen of newly born pigs began to scour and grow thin, and soon after another penful succumbed. Despite all the treatment the vets were able to offer, these piglets died, as did every pig born for the next four weeks. We were at our wits' end; antibiotics were of no avail. We tried, unsuccessfully, feeding diseased material to pregnant sows hoping they would catch the disease and build up enough antibodies before they farrowed down. We even tried milking some of the recovered sows and feeding this milk to baby pigs, but without success. We made up all sorts of concoctions, then after about four weeks some of the new-born pigs survived and then more and more until the situation was almost back to normal. The explanation appeared to be that the sows, having caught the bug, produced antibodies which they were then able to pass onto the pigs either before they were born, or maybe through the colostrum immediately after. This was not the end of our problems, however. Despite our efforts to control the spread of the disease it entered our other units with similar results.

Even with such challenges, pigs were profitable and we decided that we should expand on a new site taking advantage of some of the building grants that were then available. We had been mulling over for some time how we should go about it. The solution, as we saw it at the time, was to build another 120-sow farrow-to-finish unit on a site away from any other pigs. This would help reduce the transfer of disease from one unit to another and give us room for expansion. The site we decided on was part-way up College Lane, not too far from Red House, and was known as Jacksons because the Jackson family had been previous owners of the land.

This new development at Jacksons in 1964 would consist of four buildings and incorporate the latest thinking. I spent many evenings drawing up the plans. If successful, we might use the unit as a blueprint for further units. One of the sheds would be a dry sow yard divided into four. At that stage we still felt it essential that pregnant sows should be able to go outside onto grass, so bearing in mind the problems with the quagmire outside the previous shed after a rainy spell, facilities were made for each yard to have fewer sows and to have separate access to a paddock. A second shed would be dedicated to farrowing and fitted with modern facilities. In addition to all the features incorporated into the College Farm farrowing shed it would have glass wool insulation to the roof and gates would be fitted between the pens: when open, the pigs would have a much larger pen, but the gates could be closed to make a passage on each side of the shed wide enough for a Fergie tractor with fore-end loader to push the muck right through the shed and out onto a large concrete-based muck pad. By making the passage wide enough for a small tractor we were able to reduce the amount of mucking out to be done by hand; this was quite a breakthrough. The muck could then stay on the concrete pad until we were ready to spread it onto the land after harvest each year. Sheds three and four for the weaners and finishers respectively would be arranged for clearing out the muck in a similar way. At the top end of each of the four sheds was a bulk-meal tank which we filled using the new bulk-meal trailer to cart the meal from our grinding shed.

In the middle of that building project in 1964 our pig manager, Tony Britton, who had been with us for seven years, decided to move on. Fortunately I knew that Nigel Marshall, who had worked with the pigs at our Norton Farm, could be persuaded to come back to a more senior position. He had moved from us when offered a job as foreman on a neighbouring farm. Pleased to accept my offer, Nigel returned but this time as pig manager and it was the beginning of a very happy and fruitful partnership working together until he retired. We were both keen to see the pigs expand and we travelled a lot together as we sought to keep up with the latest developments.

I travelled with Nigel in 1964 to Northern Ireland with the FBA. Nigel always remembered, when expressing an enthusiasm for push-through fattening sheds such as we were in the middle of building, being put down by one of the senior members of the party. It was a time when the hot money had moved on to keeping pigs on slats. The attraction of slats, effectively perforated flooring, is that they reduce the need for straw and the labour involved in handling it. The pigs are separated from their dung and kept drier than on deep straw, which, theoretically anyway, should lead to a healthier animal. It was on this visit that we first saw pigs being successfully housed on slats. Widespread research was being carried out into the use of slats, much of it by the Scottish Farm Buildings Unit, and I was very interested in what could be learnt and how the system might help our pigs at home. There were many questions to be answered. What were the most suitable materials: wood, wire mesh or concrete? What was the ideal shape and how wide should the gap between slats be to suit different sizes of pig? How would the pigs react to them? There were also new issues that their use brought about, the prime one being how best to handle the slurry from under the slats. The solid farmyard manure had been little problem, heaped on a concrete slab at the pig unit. There it stayed until we cleared it after harvest using a Bray loader and three Farmhand muck spreaders, carting and spreading it on the land before ploughing.

We would now have a semi-liquid material, slurry. Ideally, we needed to use it as a fertiliser but there were only certain times of the year when we could put it on the land and, if we applied too much, it could find its way into our drainage systems and into the watercourses. This meant we needed appropriate methods for dealing with it including adequate storage. Did the channels need a slope? If so, how much of one? Could one store the slurry under the slats, or should it drain into an outside tank? One farmer we knew in Scotland had built an experimental shed in which the slurry circulated partly under the slats and partly in the open air, where it was aerated with an enormous paddle. I am not sure there ever can be a best way.

Different systems suited different farms, depending on the quantity to be spread and the soil type. Much of the slurry-handling equipment in use today was not available then. In 1969 I visited farms in the Po Valley in Italy, again with the FBA. On one farm, to my surprise, the slurry from the cows was being deliberately released into the river Po itself. On the Ferruzzi farm at Ravenna, which had 10,000 sows and over 100,000 fattening pigs, it was pumped into large, circular slotted tanks, the liquid fraction passing through the slots and the solid matter retained in the tanks being left to dry out. The liquid was pumped directly onto the surrounding vines and, after drying out in the hot, dry Italian summers, the solid material was spread onto the land in the same way as farmyard manure always had been. Back in the UK, we heard of one farmer who had permission to pump slurry into the Humber estuary on the ebb tide. Another pumped it continuously onto grassland and said that it permeated the soil without reaching the water table below. Clearly these were not going to be solutions on our heavy clay in mid-Suffolk.

We visited numerous farms where slats had been successfully installed and the slurry taken out onto the land using a slurry tanker. We decided that the system showed promise but that the best way to find out was to try it ourselves. We made the concrete slats and installed them in a new fattening building as we expanded the pig unit at Jacksons. This was the obvious choice for experimentation with a new system as it was far enough from dwellings in case a strong smell was generated. We had already put up the buildings for a 120-sow farrow-to-finish operation and there was space for development when we were able to expand. The slurry would be collected in a tank at the lower end of the fattening shed. From there it would be pumped into lagoons which we bulldozed out on land we always referred to as 'the bomb dumps', some 700 yards away. It had been part of the Great Ashfield aerodrome built for American Flying Fortresses during World War II and, away from the hangers and runways, was where the bombs were stored.

As our Jacksons pig unit grew, there were challenges, and pumping slurry was just one of them. The first pipe we used had too small a diameter. Periodically it blocked up and needed clearing, and often I would muck in (literally) and help sort out the problem. Mary was not best pleased if I returned home accompanied by a very strong smell, particularly if we had guests. I soon learnt to peel off my overalls outside the house and shower before reappearing. It was not until we put in a three-inch pipe and increased the flow rate to prevent the solids from coming out of suspension that we finally solved the problem.

That first slatted-floor fattening shed we built at Jacksons was based on one we had seen just outside Huntingdon. It had 47 pens, each capable of holding up to 17 pigs. It worked well and, once this slatted-floor building was up and running successfully, we considered using slats in other stages of pig production, practices Nigel and I had by then seen on several farms. We introduced them in dry sow stalls and farrowing pens but continued to use straw in the service area and in the multi-suckling pens. Multi-suckling was another stage we went through. Rather than take the piglets away from their mothers at just three weeks old, we brought sows with their litters out of the farrowing shed and put four sows and their litters together in a pen, allowing the young to suckle for another two weeks. When we then took the sows away, the piglets could stay in the same pen for another eight or 10 weeks until we moved them to the final finishing (fattening) stage.

We steadily expanded the Jacksons unit until it held 800 sows and 8,000 pigs in total. This obviously meant a great deal more slurry. Initially we tried to get it to dry out, rather like the Italians, so that in the autumn we could cart it onto stubble and plough it in. This was hard enough when the weather was dry but when it was wet we made too much mess. We, and others, experimented with numerous machines to try to separate out the liquid from the dry matter. The fact that it was a very corrosive material did not help and even galvanised steel would succumb after a season or two. Many machines incorporated rollers and screens but were still not very effective. Always more solids than we would like went with the liquid, and the solids, instead of piling up, would flatten out like a cow pat. Later we tried out some German screw presses. They worked well when new, but relied on a dried slurry bung to contain the input. As the machine wore, occasionally the bung would disintegrate, allowing slurry to pour out everywhere.

We managed as well as possible. The semi-solids were handled using standard muck spreaders, and with a tail gate fitted there was better control. The liquid portion was sprayed onto the land through a rain gun, but this was not the most attractive way especially if one was down wind. We did cause another problem when, on one occasion, we sprayed through some high voltage electric cables causing a supply fault on the line from Elmswell to Thetford. We were not very popular with the Eastern Electricity Board nor, I presume, with the inhabitants of Thetford. One May and June we even tried pumping unseparated slurry onto the rhubarb fields, but the rhubarb curled up and died which did not make for good family relations with my brother, Robert, who ran the horticultural side of the business. Much of the liquid fraction was irrigated onto sugar beet in June and July, the liquid helping the beet along, but the nitrogen depressed the sugar content and added detrimentally to the nitrogen levels in the harvested beet. After all these years of experimenting, our best answer so far has been to store the slurry in a lagoon and, when spreading time comes, stir it before pumping it through an umbilical pipe to a slow-moving tractor with the facilities to spread and incorporate it into the soil. This can be done not only after harvest on bare land but also onto a growing crop in spring when the nutrients are most needed.

It is said that what goes around comes around, and now in the 2020s, with pressure from our customers who like the image of pigs bedded on straw, we have moved away from slatted floors. Most of our pigs today are bred and reared either outside or on straw, albeit in a much more mechanised, less labour-intensive way than 50 years ago.

Another major development we went through in pig production was the use of liquid feeding systems, introduced at the same time as we were installing slatted floors. Here we were much indebted to John Hitchcock and his electrician, Maurice Aldous. They had spent a lot of time and experimentation developing a system, mixing meal with water or other liquids, and then pumping it through 2in pipes around the shed to the feeding troughs. Using pipeline feeding, it was easy to benefit from cheap skimmed milk when it was available. It was a very useful protein, too, and the Danes, pre-War, considered it essential. In some years, skimmed milk was available from the Stowmarket Creamery

Farm tanker and skimmed milk, 1950s and 1960s

at a reasonable price. We even bought a second-hand tanker because we were feeding up to 20,000 gallons of skimmed milk per day. Eventually the Stowmarket Creamery closed and the cost of sourcing skimmed milk from elsewhere became uneconomic. We included liquid feeding in most of our subsequent buildings, which was helpful when other liquid products became available. At times we fed C starch and, later, abrocarb, both by-products of producing sugars from wheat. We also help the brewers by using beer waste which they cannot put down the drain; it has some food value, but the quality is so variable that we only use it as a replacement for water.

Initially the mix for pipeline feeding was let into each trough using a hand-operated valve. The pigman opened it for a given amount of time depending on the number and size of the pigs in any pen. We thought the system made feeding easier, but such is the law of unintended consequences that some years later an ex-employee, advised, no doubt, by a no-win no-fee lawyer, sued us for repetitive strain injury, a claim eventually settled by our insurers. Later the valves were automated using compressed air. We controlled the timing by using second-hand Strowger dialling switchgear, another of Maurice Aldous's brainwaves. These mechanical switches, which were being replaced and sold off by the Post Office, served us very well for 25–30 years but eventually wore out and we too replaced them with a computer-controlled system.

Control wires and compressed air pipes led to a valve for each pen, although we were able to reduce the number of valves needed by siting a partition wall over the trough between two pens, allowing pigs to feed from each side. This enabled us to feed 34 rather than 17 pigs from each valve. As always when embracing new technology different problems emerged. Vermin were a problem. The long lengths of wire and compressed air pipes to each valve were particularly prone to damage by rats or mice. We did our best to control them but occasionally we would get a pipe gnawed into or a wire bitten through, leading to food going into the wrong trough or overflowing onto the slats. It was not only food flow that they could disrupt. One night we had a control wire bitten through in a flat deck room holding 200 pigs. The controls thought the pigs were too cold and continued heating the room. In the morning, sadly all 200 pigs had died with heat stroke. It was an expensive way to find out that we needed better fail-safe dropouts on the ventilators. Another pest we encountered was the dermestid beetle. These beetles would get into electric trunking and seemed to be partial to the plastic-coating on the copper wires, causing short circuits, trips going out and the possibility of an electrically induced fire, though fortunately we never had one.

Feed is 70–75 per cent of the cost of producing pigs. Until 2017 we milled and mixed all our own food which over the years gave us considerable flexibility. Much of it was our own wheat and barley but, when it was more profitable to sell the grain and replace it with tapioca, maize, rye or potato meal, we did so. However, in 2017 we stopped producing our own rations. The main reason was because most of our pigs are now sold on a contract largely dictated by Waitrose, which demands uniformity among their suppliers as to production methods, genetic sourcing and feed rations. Additionally, our existing pipeline systems were showing signs of wear and needed updating. After studying a raft of pros and cons and with much heartfelt consideration, we made the decision to stop mixing our own feed and to buy it in from one of the national feed companies. We have, however, maintained some cost control by doing a deal with this feed supplier to use our grain when appropriate.

The 1970s, 1980s and 1990s were years of continuous expansion of our pig enterprise. As well as the farrow-to-finish unit holding 800 sows at Jacksons, we built similar sized units at Hardings Farm in Norton, Abbey Farm in Wickham Skeith and Scholes, near the bomb dumps at Great Ashfield. No two units were exactly alike for this was a period when new ideas and techniques were constantly being explored and we were keen to keep up with new developments. We managed to keep the cost of building these units to what we felt was an appropriate, not overly expensive, level. The complete unit at Hardings Farm was built for us by Aves[3], a local building firm, in 1974 for £152,301. By then we had sorted out the slurry-handling problems sufficiently so that most of the buildings on these units could be slurry based. The expansion was such that in 1982 we sold 46,127 pigs.

We no longer needed to make our own slats as contractors were by then manufacturing them and we could pick and choose from whom to buy. However, concrete was not the answer for all sizes of pig. We were approached in the early 1980s by Peter Leveret who had an extrusion plant for making plastic slats. He claimed they were more suited for weaner pigs and farrowing pens than concrete slats, as they had a greater percentage void than one could get with concrete. Nigel and I thought they had potential and decided to incorporate them in what we were building at the time. Unfortunately, sometime later Peter ran into financial difficulties, but we were now committed to installing these plastic slats. What were we to do? We agreed to buy the extruding equipment and continue producing for our own needs. As luck would have it, Phil Walker, a relative by marriage who had run a powder coating business, happened to be looking for a new job. He came in to run the extrusion plant and both Les Gooderham and Andrew Cuthbert from our workshop lent a hand. We extruded enough to be able to install them in farrowing pens and flat decks at all the farms (Norton, Jacksons, Scholes and Wickham Skeith). Eventually they were superseded by a different style of plastic slat – injection moulded – and produced in large quantities on the continent.

In these later developments, instead of using multi-suckling as we had at Jacksons, we built temperature-controlled, fully

---

[3] Aves builders: Jack and his son Albert Aves built many of our pig sheds and farmhouses.

slatted weaner accommodation, known as 'flat decks'. Each room or flat deck could hold up to 360 pigs in pens of 15. Fans blew air through plastic tubes running the length of the room to keep temperatures stable, and shutters were incorporated to enable us to introduce fresh air as required. Much of the planning was in collaboration with Geoff Owen and Hugh Crabtree from the University of Reading, where they had been studying ventilation in livestock buildings. Together they went on to found Farm X, a company which has specialised in developing equipment for automatic climate control in buildings as well as other computer-controlled systems it markets on a global scale. We were one of their early customers and I was offered an opportunity to see their ideas rolled out in a commercial setting at a dairy farm in Somerset. In those days, the route to Reading and the West Country was via London, and I well remember driving past Big Ben at 6am and then again at 11 that night on my way home.

Tommy Wright mixing meal in the mill, 1960

Farrowing shed of the 1950s

Kenny Betts carting out pig meal from our grinding
shed (mill) on home-made go-cart

With Father in front of the grinding shed, 1976;
custom-made meal carts can be seen behind us

Fattening pigs in this way into the 1970s

This pig fattening shed is typical of those we were building
throughout the 1960s and 1970s

Flat decks at our Norton pig unit, 1976

Indoor farrowing pens of the 1970s and 1980s.
Note the bars to prevent the sow from crushing her piglets

Aerial view of our pig unit at Hardings Farm,
Norton, taken in 2011

So much for the pigs' housing, but what about the pigs themselves? The animals we deal with today differ considerably from the animal we were producing when I began farming 70 years ago. Looked at through today's eyes some things from that period look strange. For example, it was then not permissible to use an unlicensed boar, and a boar could not be licensed unless it was a pure bred and its pedigree could be shown. Even then it had to be inspected and approved by a livestock official from the Ministry of Agriculture. They thought this 'control' was a good idea at the time, but it severely limited the genetic development of the industry. Others must have thought so too for, in the mid-1940s, the Animal Breeding Research Organisation (ABRO), a branch of the University of Edinburgh, had been set up to investigate the genetics and practices of breeding farm animals. Change came slowly and, not surprisingly, was opposed by the various breed societies. ABRO, now known as The Roslin Institute, was the organisation that many years later produced Dolly the sheep, the first cloned animal.

I have said previously that after the War there was a feeling that the Scandinavians had a better pig than we had in the UK and in 1949, after some intense lobbying, three groups of Landrace pigs, after passing through a quarantine period, were imported into the UK from Sweden. Each group comprised three or four boars and eight or nine gilts[4]. This original importation of Landrace pigs from Sweden was permitted on the basis that the progeny should stay on the original farms until the Ministry of Agriculture was able to carry out an evaluation of their worth to the UK pig industry. My trip to Yorkshire and Scotland with Stephen Horvat in 1949 had enabled me to see two of these three groups and I had come home bubbling with enthusiasm for the Landrace breed. It was two years later that the Ministry of Agriculture declared that these imported pigs and their progeny were

---

[4] Gilt: a female pig who has not yet had a litter of piglets.

showing promise and that the government would lift the restrictions on their sale to other farms. At Bacton we were keen to benefit from cross-breeding our Large White pig with the Landrace, improving the quality of bacon we could produce, and in September 1951, within minutes of Grandad hearing that Henry Wertheimer had some Landrace boars for sale, he and Father were in the car on their way to Yorkshire. By lunchtime they had bought two of them. Unfortunately, only a few months later, one of these boars escaped from its pen and found its way into a meadow in which some sows were grazing in the company of a Large White boar. The Large White set about the Landrace boar and killed it. Fortunately, the other went on to father many Large White/Landrace crosses which were well received when they went to the factory.

In 1953 the importation of a further 25 Landrace boars and 75 Landrace gilts from Sweden was sanctioned by the government. They were to be auctioned at Peterborough in May that year, each lot comprising one boar and three gilts. Nobody really knew what they would fetch. Stephen Horvat heard that Mr Crease, who ran the Bluegate herd for Strutt and Parker Farms in Essex, was wanting to sell 12 in-pig gilts, pure bred from the original Landrace imports, to fund the purchase of a group of the new imports at the auction. So a week before the Peterborough auction I went with Father, Grandfather and Stephen to see these gilts. Mr Crease was asking 180 guineas each. Stephen was all for buying the lot. Grandad was more cautious: fat pigs were then fetching £18 and he felt that 180 guineas was therefore a high price per gilt. After much wrangling, it was finally agreed that we would buy one gilt and Stephen would buy another. How wrong can you be? This was an expensive error. Just one week later, those same pigs would have fetched £1,500 each, such was the demand at the Peterborough sale. Stephen later said, 'You should never go shopping with someone who can afford to make a mistake.'

Following on from the sale at Peterborough, a Landrace Breed Society was formed which registered births and pedigree for these pigs and their progeny. Stephen was one of the founder members. We were members for a time and continued to take an interest in the Landrace developments, slowly increasing the number in our herd. There was a good demand for breeding pigs with the Landrace genes as the cross-bred hybrid sow produced on average an extra half a pig per litter, a stronger piglet at birth and one which grew to slaughter weight marginally faster and with less backfat. We turned the Abbey farm site at Wickham Skeith over to producing these cross-bred gilts, initially to supply ourselves with breeding stock for use on our other farms. When we produced more than we needed we sold them on. Mostly they went to United Pig Breeders (UPB), a company set up by a group of local pig farmers specifically to sell such breeding stock.

UPB developed connections all over the UK and even had an export business to Holland. I went there with one load of pigs as companion to Trevor Owen, chairman of the group. We travelled overnight on the ferry from Harwich to Hook of Holland and had four deliveries to make. The first we made in time for breakfast and a drop of Jeneva, the local gin. I see from notes I made at the time that at the subsequent farms we were offered coffee and more gin; the Dutch farmers are a very hospitable lot. Our deliveries were all to smallish family farms, with herds of no more than 80 sows, the houses bedecked with flowers and plaques declaiming such wisdom as 'To do first and think later leads to much grief', 'Enjoy your life for it is short' and 'Many people have too much but nobody has enough'. Like most people, I had heard that the Dutch traditionally wore wooden clogs but was surprised that most of the farmers who took us round to see their pigs really did slip on a pair of wooden clogs.

Gradually, throughout the 1970s, the traditional butcher's and grocer's shops were being squeezed out of business. The supermarket was on the rise and our connection with the bacon factory at Elmswell gave us an insight into the way they were thinking and developing. Elmswell had seen the way trade was going and had begun pre-packaging more of its products, rebranding these in 1971 as Farm Kitchen Foods. It had also developed links with Marks & Spencer (M&S) and was producing items to their specifications.

By 1980, it was becoming apparent that the concentration on producing leaner pigs which we had achieved with the importation of the Landrace breed was leading to a less juicy pork chop. M&S was anxious that something should be done about it as they had a feeling that the Danes and the North Americans were producing a better product. Peel Holroyd, their chicken and pork specialist, was sent to Denmark and North America to sample pork chops and to investigate the subject. He left a trail of half-eaten pork chops before returning and reporting back on the housing, feeding and breeding practices in each area. These eating trials pointed to the presence of more interstitial fat or 'marbling' in the meat, particularly associated with the Duroc breed. The next step was to test it out.

Peel approached us to run trials to test his theory. We managed to find some cross-bred Duroc pigs which we put onto the Mendlesham unit, weighing them, checking the food conversion ratio and so on. When they eventually arrived at the bacon factory the chops from them were cooked and compared in eating trials with chops from our general run of Landrace/Large Whites. The results were such that M&S asked us to breed more of these Duroc-cross pigs. So it was that in 1985, partly paid for by M&S, I travelled to Denmark together with Nigel Marshall and David Neal, general manager of the Elmswell bacon factory. By then the Danes were prepared to consider the export of breeding stock and were very helpful in showing us some of their established Duroc herds. Peel and I chose Denmark rather than North America because there would be fewer problems importing animals from there. We also knew that the Danes, having imported Durocs themselves from Canada some years previously, had kept them under close supervision and had a large amount of data on their performance. We bought six Duroc boars and six Duroc gilts which, after passing through quarantine, came onto a unit we hired from the Luxfords at Old Newton Hall.

Initially we were using natural mating which was not producing pigs at a rate fast enough for M&S. To overcome that, in 1988 we set up a semen collecting unit at the old College Farm in Bacton, starting with half a dozen Duroc boars and the necessary laboratory equipment. Paul Braybrook, in the farm workshop, made an artificial sow out of metal which served a useful purpose. From one ejaculation, between 20 and 50 sows could be impregnated, depending on the liveliness of the sperm, which was examined under a microscope to determine its motility and viscosity. My son, William, set up and ran the artificial insemination (AI) operation and trained up one of our younger pigmen, Lee Potter. We received much help and guidance from Dr Christianne Glossop, a local vet who did much to encourage the development of AI in pigs throughout the country. Later, in 2005, she was appointed chief veterinary officer for Wales where she now spends more time combating TB in cattle. AI is now well established and licensed. The development of genetics and the use of the computer have speeded up improvements in this and other areas of pig production. To take advantage of these improvements and to reduce costs, pig producers no longer need to keep boars and most now buy in semen from one of a number of specialist breeders who employ their own geneticist and reap the benefits of scale. In time, we too found it to our advantage to buy in the semen and so closed our own production unit.

But to return to the situation 40 years ago. Once we had produced a number of Duroc-crosses to bacon weight, the factory ran extensive pork-chop eating trials. Eventually we were able to provide M&S with what they wanted, namely, juicy pork chops and bacon rashers from an exclusive cross using Durocs as the terminal sire[5]. In an article about the Elmswell bacon factory in the *East Anglian Daily Times* in March 1985, Peter Hopper, the newspaper's agricultural editor, recognised some of the challenges:

Breeding for leaner bacon

The effect of consumer preference for leaner bacon has gone back through the food chain to the farmer, who has geared himself to producing the type of pig that will give us meaty rashers. But changes that must start with the genetic 'improvement' of a living species do not happen overnight. In fact, many years of painstaking breeding work is involved. While medical bodies may have sound reasons for deciding that consumption of animal fat is bad for health — it has been linked with heart disease — the farmer has to be sure that it is not merely a whim of the time. Fashions in food change like everything else.

Mr David Black, whose family farming company at Bacton is the largest supplier of pigs to the Elmswell factory, urges some caution in the pursuit of leanness. He says pig producers have been striving for many years to get leaner bacon pigs, but in doing so they have to be careful not to affect the succulence or eating quality of the bacon. Pressure from the supermarkets for processors to supply lean bacon to their own tight specifications is rigidly applied and that involves the farmer constantly reviewing techniques in breeding and feeding.

In some ways we became victims of our own success. M&S was pleased with the loins and the pork chops for which it was prepared to pay a premium, but the remaining cuts tended to be fatter than the general trade required and had to be sold at a discount. While all our production was going to M&S this worked out well, but increasingly, despite

---

[5] Terminal sire: 'terminal' as we were not intending to use progeny from this sire/boar for breeding.

asking us to keep producing more loins and chops, they would not always take them all, nor every part of the pig. It could have been made to work if M&S had asked manufacturers of their pork pies, sausages and Scotch eggs, for example, to take other parts of the carcass, but it was not prepared to do so. We lost out financially on the animals we had bred to the company's specification, and there were progressively more of them. Negotiations with M&S failed to solve the problem and eventually we gave up the programme.

M&S was also interested in the possibilities of selling meat from outdoor-bred pigs, and we worked with them from 1987 to develop outdoor pig production, initially in association with Peninsular Pigs, a UK company set up in 1977 especially for the purpose. Until then our focus had been on producing pigs as economically as possible in indoor units. We were approached by Eric Oakes of Peninsular Pigs with a deal to produce weaners farrowed in arcs outside and finished on an all-straw-based system. We had already seen such a system run by Richard Roadknight in Oxfordshire. Concerned about farrowing outside in the winter, his sows only farrowed in September and March. He was using a Blue and White cross because it was deemed more robust for the conditions even though it produced a carcass which was over-fat. Eric assured us that he had a strain of pigs every bit as suitable for outdoor production as those which Richard Roadknight kept, and that they would produce carcasses of a much better quality which would be acceptable to M&S.

We owned no land which was sufficiently free draining for running pigs outside, but Eric had located light land fit for the purpose which we could rent from the Pauls at Broxtead Farm near Woodbridge. He was keen for us to get involved, based on our earlier work with M&S and our reputation in the pig world. After studying the possibilities we agreed to give it a go and so, starting with 100 'Peninsular' sows we embarked on yet another project. Although Eric had set up a number of units, we soon found ourselves in all sorts of unfamiliar territory, needing answers to wide-ranging concerns: electric fences to keep the pigs in and the foxes out, design of the farrowing arcs, accommodation for the dry sows, what to do when the frost froze the water pipes, finding and training staff, how best to gather up the pigs when it became time to wean them, what type of carry-on housing. As ever, we experimented and steadily expanded, though we did have other problems such as when gale force winds blew some of the farrowing arcs into the hedge and when we suffered arson attacks on stacks of straw. Over the following years, we and many others built up experience and some expertise in this alternative way of pig keeping.

We continued to have pigs at Sutton until May 2002 when the Pauls wanted the land for vegetable production. We moved production initially to Alderton also on light land near the River Deben, then to Euston, near Thetford and then to Herringswell, near Newmarket, on land belonging to Uptons. While we were on the Euston estate, we had an unfortunate experience on one August bank holiday weekend. It had been particularly dry and the electric fencing was not working as it should. Fifteen hundred little pigs decided it would be a good idea to go exploring and found their way out of the field into the parkland. I received a telephone call from a very irate agent saying that the Duke of Grafton was much displeased with the situation and would I sort it out. Nigel and I, together with the pigmen, spent the rest of the weekend trying to gather up as many of the escapees that we could and transfer them to some hastily arranged accommodation elsewhere. It was shortly after that incident that we were requested to move on, our pigs no longer welcome on that site.

At Herringswell we built the herd up to 1,500 sows and housed them in arcs made from cutting second-hand 13,000-gallon water tanks into four. These are not only cheaper than the corrugated iron arcs currently on the market but will last longer and are heavy enough to prevent the wind blowing them away. It means that a teleporter is necessary to move them but, since one is required on the unit anyway for straw handling, it is not a problem. We still have pigs there today, but the herd is reduced to 600 sows, as much of the land has been taken up with providing sand and shingle for the ever-expanding township of Red Lodge. At present, these outdoor piglets are weaned at three and a half to four weeks old and moved to Hardings Farm where they are finished on straw.

There are now, in the eastern counties, so many outdoor weaners being bred that it suits us better to buy in the extra piglets we need to fill the fattening (finishing) capacity at our other units, rather than to rent more light land to produce them ourselves. We buy in approximately 800 of these outdoor-bred weaners from British Quality Pigs (BQP) and rear them outdoors with tents for shelter and beds of straw on other light land, currently rented from Rougham Estates. When they are 12 weeks old they are moved into our straw-based finishing units. After each batch of pigs, the tents are moved to prevent the area becoming pig sick. Rougham Estates is keen to do a 'straw for muck' exchange which works well for both parties, the muck being a useful source of organic matter plus nitrate and phosphate for its lighter sandy soil. The tents have a lifespan of about 10 years but such is the cost that the housing is much cheaper than building something more permanent.

Since 2013 most of our production has gone to Waitrose and it is gratifying that the company appreciates that a pig has four legs, a head a belly and a loin and that the whole pig needs to be sold in balance. In exchange for producing pigs according to Waitrose's specification we have a guaranteed premium. Among the supermarkets, we believe that Waitrose now has, in the public's perception, the top-spot for quality meat. However, with the cut-throat competition there is in the retailing business, who can say how long that will last?

Thinking about pig production in general, back in 1980 I was asked to speak to a group about intensive pig farming and the welfare lobby. I had seen a programme on television, and there have been many more since, heavily angled against

Duroc pig introduced to breed a 'better' pig

Checking a piglet in outdoor arc, 1992

Pig arcs as used in our outdoor units, 2018

Weaned piglets in tents on outdoor unit at Ixworth, 2018

Alternative accommodation for weaned piglets
on outdoor units, Cosy Kennels

Teleporter (fore end loader) used for loading straw
and moving pig arcs on outdoor units

intensive production, showing animals in idyllic outdoor conditions compared with the worst of the indoor ones. It would have been just as easy to contrast outdoor animals suffering in the mud, slush and cold of winter-time conditions with the comfort and contentment of the better indoor units. There are those who complain genuinely about the ill treatment of animals, and I am sure we would all go along with them. But it is the way the animals are managed rather than the fact that they are being kept extensively or intensively which, assuming one is anthropomorphic, determines their quality of life.

However, largely due to intensive lobbying and the public's perception of what is 'good', more and more pig keepers have embraced the development of breeding and rearing pigs outdoors, such that much of the light land in the eastern counties not being used for intensive vegetables supports an outdoor pig herd. Along with outdoor farrowing, which is very weather dependent, there has been pressure on pig keepers to finish their pigs in straw-based systems rather than on slatted floors. This means that the cost of production has gone up, but if this is what the public is prepared to pay for, then this is the way the industry will adapt and develop. Pig farming has gone through cycles, from straw-based systems to slats and stalls, and now back to straw-based systems, which are considerably more labour intensive and weather dependent. Straw may look and seem more appealing, but the muck and urine are not removed instantaneously as they are through slats. Slats keep working even when the lavatory attendant is not there.

A pig producer has a range of factors to juggle, some of them conflicting, to produce pig meat profitably. A profit must be had, or we should not stay in business long. We want to see our pigs well housed, well managed and well fed. Similarly, it is important to look after our staff and to ensure they are well rewarded. To do this, one must have the appropriate price for the finished pig. One thing is certain: there is little stability in the pig industry. That is as true today as when I wrote an article for the Rank Hovis McDougal newspaper *On the Farm* in 1975, saying, *The price of feed continues to fluctuate, and pig prices remain volatile. The pressures on our pig industry to get on or get out are intensifying. What I can see is that herds must continue to expand so that they can afford to employ specialists, yet not so big as to be beyond the supervision of the boss.*

What the optimum size for a pig herd is, I do not know. I have been to pig farms in the US and Russia, where the numbers are mind-boggling; some have 5,000 sows and their progeny and are intending to increase that number to 25,000 sows plus offspring. That would make a total of 250,000 pigs on one unit. When I expressed concern regarding the potential for the spread of disease I was assured that with tight security one could maintain high health standards. There were food production units, slaughtering and processing facilities on site. With the sheer amount of traffic bringing in feed and taking out produce and handling effluent, to say nothing of the number of staff required, I would be very concerned about biosecurity.

When everything goes well there are efficiencies of scale and, with vertical integration, the cost of pig meat can come down enormously, as has been seen with the production of chicken meat. The risk factor is high, however, and we have yet to see how the public and governments will react to food brought in from such enormous operations as these. Animal welfare standards are rightly imposed on production in the UK but farming to high welfare standards is more costly for the producer, and it is crucial that the market is not undercut. Imported meat must be of the same standard. Whatever happens it is not right that the UK pig industry should be subjected to legislation that adds to costs, and then be expected to compete with lower-cost imports which are not produced to those standards.

There is another sobering thought that comes from globalisation. It is often difficult to know who the ultimate owner of a particular company is. Currently we are closely involved with BQP, buying weaners from them and selling finished pigs through them, reared to the Waitrose specification. BQP was a subsidiary of a Danish-owned operation called Tulip but recently we heard that Tulip had sold BQP to the American meat firm Pilgrim, which in turn is part of a much larger meat business, JBS, based in Brazil. It is rather disconcerting to suddenly find that one's pig business could be seriously affected by a decision taken in Rio and even linked to a firm which might be responsible for damaging the rainforest.

CHAPTER 7

FARM BUILDINGS

In 1900 all Suffolk farms had a stackyard, with a barn and barn yard for fattening cattle and maybe a few pigs running with them. Close by would be the stable, where the corn was stored on the first floor. It was said that you weren't a man until you could carry a coomb[1] sack of beans across your shoulders up the granary steps. At Red House the old granary is the only original building to have survived, though woodworm and beetle have wrought havoc with the woodwork. It had a thatched roof, but this was replaced with corrugated, galvanised iron in the 1940s and all the other buildings have been replaced.

When Grandad and Grandma arrived from Scotland with their Ayrshire milking cows, Grandad persuaded the landlord to build a cow shed. That cow shed has long since gone but the area in which it stood is still sometimes called the nettus[2]. There would also have been several other ramshackle buildings in which chickens could be shut up at night to prevent them falling prey to local foxes. There were no specialist pig buildings until the late 1920s, though even these were simple wooden sheds with a corrugated iron roof, sometimes without even a concrete floor.

Wheat and barley were not very profitable in the 1930s and so the farm diversified into many different crops. Sugar

---

[1] A coomb sack of beans weighed 19st or c.120kg. A coomb was a measure of volume hence a coomb sack of beans weighed 19st, wheat 18st and barley 16st.
[2] Nettus: a house or shed in which cattle are kept, derived from 'neat-house'.

beet, a labour-intensive crop, was actively being sponsored by the government to help with the national unemployment situation linked with the Great Depression. I was told that for several years, men would cycle out from Stowmarket to hoe and then later to pull, knock and top beet. A wide variety of fruit, vegetables and flowers was also grown, with some sold locally but most going to the London market at Covent Garden.

Apples sold directly after picking were not very profitable, but it was a different matter if they could be stored and eased onto the market. A shed was duly built to store these apples in bushel boxes. It had a cellar, because it was thought cooler below ground and that the temperature would fluctuate less. It could not have been very satisfactory because refrigerated cold stores were installed in 1936. An extension was added to the apple shed to provide more room for the flower-packing operation and this, suitably decorated with flags and bunting, provided the venue for a tea party for the village children to celebrate the Coronation of King George VI.

Little new building or maintenance was done during World War II. Afterwards, however, there was a pent-up need for new buildings to accommodate the changes that were taking place. The combine harvester was replacing the binder and threshing machine, bulk handling replacing sacks, tractors and their servicing requirements replacing the horse. About a quarter of the land which we farmed had been needed to provide sufficient food and grazing for the horses, land which was now available to produce crops. This, together with new higher-yielding varieties of cereals, meant we had a need for more storage facilities.

There were problems enough in a wet harvest when corn was cut with a binder and put into stacks, but with corn in bulk, getting it dry enough to store became more important than ever. So, during the winter of 1961 to 1962 we installed a Penny and Porter drier with a paraffin-fuelled furnace in the grinding shed. We had a need to dry gladioli corms as well so, ever conscience of costs, had the burner mounted on wheels so that it could be moved to do either job. We were just in time, because 1963 was a wet harvest and we were combining wheat at up to 25 per cent moisture. It was a challenge to get this dry enough to store. If not sufficiently dry it would begin to go mouldy and a whole silo full could be spoilt. To increase capacity and the rate at which we could put the grain through the drier, we removed the cooling part from the drier and put the warm grain, which we had managed to reduce to 20 per cent moisture, straight into the silos. To reduce the temperature and provide the necessary extra drying we installed a facility to blow fresh air through the grain mass. It was up to 15 feet deep, something not recommended, but it worked, although we had to put in a fan to remove the condensation which formed on the underside of the roof. In this way we were able to reduce the moisture content by a further 3 to 4 per cent.

We contacted Mr Hollinrake, the Ministry of Agriculture's building advisor based in Cambridge, and Mr Brunton, who ran a company designing and erecting steel-framed farm buildings. At six shillings and eight pence a square foot (equivalent to 3 sq feet for £1) they were good value and we put up five such steel-framed buildings in the 1960s and 1970s, each 90 feet wide and up to 105 feet long to replace the old barns on the various farms, providing much more convenient storage for grain, fertiliser and machinery.

The 'top shed' in the further stackyard at Red House was the first of these post-war buildings. Protein for animal feed was in short supply; this was the time of the groundnut fiasco. The UK government launched a scheme to develop three million acres in Tanganyika with the intention of producing groundnuts (peanuts) to fill this shortage. The politicians pushed it through in great haste, but it was a disaster from the start. Nobody had checked the soil suitability, rainfall, infrastructure or the possibility of cash being syphoned off in payoffs and graft. Like so many of the government's grandiose schemes it stumbled on before finally being abandoned.

Closer to home, grant aid was being offered towards the cost of producing dried grass and lucerne meal which could contain up to 25 per cent protein, and we drilled some 15 acres of lucerne, or 'alfalfa' as the Americans called it. With our own staff, we put in shuttered concrete walls and installed a Templewood grass drier in this new shed. I recall treating the seed with nitrate-fixing bacteria which would form nodules on the roots in the same way that bacteria formed nodules on the roots of other leguminous plants such as peas and beans. Apparently the bacteria are quite choosy and, never having grown lucerne on the farm before, we had to ensure that the proper strain was available. The seed arrived with a culture which I had to mix with skimmed milk, thoroughly incorporate it with the seed, then allow it to dry off before it was ready to drill. The crop grew well and in time we were able to harvest three cuts a year which we dried and ground, giving us a rather dusty, green coloured, 20 per cent-protein meal which we incorporated into our pig rations. This project lasted for about 10 years until, with rising costs and cheaper alternatives, the operation stopped. An interesting side note to this was that we were losing pig feed from the Wickham farm. It was apparent that someone was stealing it and we had our suspicions. The police went round to the suspect who swore blind that he had bought the meal found on his premises, but he was unable to explain how it was that the meal was coloured green.

We heard of another scam at this time, involving government money, which took place at Glenrothes, a town in Fife which had been designated a New Town. New industries were being sought to provide employment for the new residents, including a coal mine which opened but had to be abandoned shortly afterwards because of water problems. An operation named Cadco, with dubious credentials, suggested that if it were provided with enough money, it could build a pig production unit with an accompanying slaughterhouse and employ 2,000 men. Cadco received the funding from the government and began some building work on the project, but salaries were inflated, new companies were formed and the money steadily disappeared. Yet more cash was poured into the scheme before it eventually went into liquidation. The principals appeared to have got away with it as there were no prosecutions. Sadly, although the buildings were offered for hire to a pig business close by, they were unable to agree terms with the town council and so as far as I know it was never used. It is so sad when one sees government (taxpayers') money frittered away on such grand but ill-considered schemes.

Back to our top shed and by the time we stopped drying lucerne for pig feed, we needed space to handle the grass seeds which we were growing for McGill & Smith. The seed came into the shed from the threshing machine or the combine harvester in sacks and most needed drying to a greater or lesser extent. Previously the seed had been put out onto tarpaulins to dry in the sun, but we now built a self-designed, in-sack drier. After drying, the seed was elevated through an aspirator to take out the chaff and very small seeds then put through a sieve to take out any unthreshed heads before being bagged again ready for dispatch to Ayr. Until Mr Beeching's cuts we loaded these sacks into railway wagons at Finningham station, but after them they had to go by road.

Another of the post-war steel-frame barns we erected was fitted with a wooden perforated floor for drying onions. When we stopped growing onions it became ideal for drying, conditioning and storing rape seed. Yet another was used as a packing shed for the cut flowers at Red House. These steel-framed buildings are still giving useful service today, albeit with some modifications internally.

In the seventies and eighties we were forcing rhubarb. It meant work for the staff after sugar beet and before we started the spring work in the fields. Using a variety of buildings gave us about an acre of indoor, insulated space, some of which was in second-hand, aluminium-roofed chicken sheds which we had acquired by offering to clear a

Monty Mutimer and mate mixing concrete for building work
at Hardings Farm Norton, 1961

Building pig sheds at Norton, 1961

Erecting steel frame shed based on Hollinrake design, 1960s.
The old barn is visible in the background, left of image

Pulling down one of the old sheds at Red House Farm, c. 1960

Sieving cocksfoot through homemade separator in Top Shed, 1961

site. In a wet winter or when the roots were frozen it could be a hard job getting the rhubarb roots in from the fields and placing them under cover. We used tractors with a buck rake on the back. By forcing the rhubarb indoors from December to March we produced a welcome addition to the nation's diet in the days before the supermarkets began bringing in fruit from all over the world all the year round. It was a very profitable crop for a while, even though we had to heat the sheds. Eventually, the costs exceeded the return and when we streamlined the business to reduce the number of enterprises and concentrate on fewer, more profitable lines, it went, along with the flowers and blackcurrants.

As we have simplified our farming operation some farm buildings have become surplus to our needs. Applying for a change of use from agricultural to light industrial use has enabled us to rent out some of them commercially to a variety of small businesses once essential adaptations and

New grainstore and workshop completed in 2017

improvements have been made. Even some of the hard standing around the farmyard is rented out, with Portable Space being the largest tenant. In 2015 the decision was taken to stop milling and mixing our own feed rations, the last trailer load leaving our grinding shed in 2017. We built a new grain store, workshop, spray storage and sprayer filling area, and knocked down the grinding shed with its associated silos at Red House. This work was completed in 2018 and redundant buildings tidied up and added to our commercial property portfolio. The board made the decision in 2020 to develop a Bacton Business Park erecting purpose-built units on the land to the south of the farm office and work on that has begun.

As well as the agricultural buildings we also owned several cottages. During World War II little was done to them and some were very tatty indeed. We had men returning from the War willing to work for us, wanting to get married and needing somewhere to live. Building materials were in short supply but we managed to get a sufficient allocation to build some new houses and add a kitchen extension to Somerville.

In 1949 we built a pair of cottages on Rectory Road, Bacton. These housed Billy Alexander, our horticultural foreman, and Horace Bloom, lorry driver and jack of all trades, with his wife Doris who ran the flower packhouse. Prior to getting married, Doris and her sister would cycle the four miles to work from Rob Hall Corner in Finningham come rain or shine. The gladioli, daffodils, tulips, irises and

White Houses built in 1948 for Bill Alexander and Horace Bloom, Rectory Road, Bacton

Wooden flower boxes arriving on Sleigh Brothers' lorry, ready to be filled with either daffodils, tulips, gladioli or peonies

peonies would be cut in the field by a gang of men and women, laid flat in large wooden boxes and brought into the packing shed stacked on the back of large, flat-trailered go-carts, to be bunched and packed. Initially, the flowers were packed into returnable wooden boxes, but later we used non-returnable cardboard ones. The boxed flowers were collected by Philip Sleigh's transport business. He and his brother ran several lorries and would collect fruit, vegetables and flowers from various farmers in the region, to be delivered overnight to the London markets.

The flowers we grew went to William Newton at Covent Garden, and the cabbages, sprouts and other vegetables to A. J. Edwards. AJ, as he was known, farmed at Bardwell. He originated from the Fen country and grew a sizeable acreage of potatoes. Not being happy with the price he was getting, he bought a stand at Covent Garden and set up a business there. Looking to expand, he approached Grandad, offering some sprout seed. He also offered to sell the produce on his stall but his main motive, he explained, was that if Grandad grew sprouts other farmers would follow suit and the acreage of sprouts grown in Suffolk would increase. We began growing sprouts and, as AJ had predicted, so did several other local farmers.

We always ran a labour-intensive operation, across a wide variety of crops and in the expanding pig business. Encouraged by grant aid from the local council in the 1950s and an ability to write off the capital value for tax purposes over 10 years, we reckoned it worthwhile to build new houses to attract and keep good workers. We built two houses at Norton Farm and a pair at Earls Green, Bacton in 1955. Frank Johnson moved into one side of the pair at Earls Green. Frank had been an evacuee from London during the War and never went back. He enjoyed the rural way of life and came to work for us precisely because we were able to house him and his bride Marian, a Bacton girl. He worked with us until he retired and later, his son Trevor worked on the farm for many years. Les Forsdyke, number two on the horticultural side, with his wife Flo moved next door to Frank with their five children, giving them more room than they had had in a two-bedroomed house in Rectory Road. It was their eldest daughter, Ruth, who became a much-appreciated Mother's help to Mary while our children were young.

We tidied up and modernised some of the old farmhouses and cottages. College Farm had been owned by Gerry Leggett until we bought it in 1953. Gerry's daughter married Freddie Whatling, who had succeeded his father as head horseman at Red House Farm, and they started married life in part of College Farmhouse, with one room downstairs and a first-floor bedroom accessed by a ladder. We modernised the house for them and they appreciated the improvements, but it always seemed to need money spending on it. However, it was important that Freddie lived close by the farm and so was worth the investment. The head horseman's job was to get the horses in at 5.30 each morning, bait them, pop home for a bite of breakfast, go back again to meet with the other horsemen at 6.30, groom the horses, saddle up and be ready for work at 7 o'clock. This was not much fun in the winter, the only light in the early hours being moonlight and a hurricane lantern. In the War years, with single summertime in the winter, it would have been even worse.

However, we considered some of the old houses to be too far gone to warrant spending money on repairs. New Barn Farmhouse was one such. We had acquired it when we bought New Barn Farm from Joe Dennis, who continued to live in the farmhouse until he died, after which we pulled it down. Since we were keeping pigs there and needed to house staff, we replaced it with a pair of cottages near Flint Cottage corner, which had the added advantage of linking into the Wyverstone village sewer system and having mains water. To give a flavour of the times, these were built by Jack Aves, a local builder, for a total cost of £2,793 including the land. Another pair of houses was built at Earls Green in 1959, which housed Kenny Betts and Monty Mutimer, workshop mechanic and lorry driver respectively, and cost a little less as we already owned the land. One of the old cottages on Earls Green was made of clay lump and thatch and housed Squiggler and his family. No one lived there after he died, the rain got in and slowly the whole building mouldered into the ground. Something similar must have happened to many such cottages all over Suffolk.

When I became engaged in 1956 I worked with an architect on plans for a house to be built at the top of the Walk on the Show Meadow. Grandfather, presumably conscious of the 1920s and 1930s, was constantly telling me not to build too big a house. 'Don't go building a palace', he said. It was a mistake to have paid too much attention to him since Mary and I went on to have four children and needed to extend the house considerably. Jack Aves started to build Harpsfield, named after Mary's Grandfather's house at Hatfield, for us in February 1957 and it was ready to move into after our wedding in September. Brother Robert married three months after us. He too managed to get Jack Aves to build a house for him and Miriam, ready for their return from their honeymoon just before Christmas. They called their home, Newyearfield, after Grandad's childhood home in Scotland. Miriam was the daughter of Harold and Hilda Cawston, who farmed at Rede near Bury St Edmunds.

Three years later we moved the farm office out of Red House into a new building conveniently placed in the farmyard. Grandma had passed away in 1954 just before I went to the States and, when Grandad died in 1960, aged 85, the old house lay empty. The books were all dispersed among the family and the furniture sold. It then seemed a ghostly place, with spiders and memories lurking everywhere. Then one day in 1961 old Red House was burnt down in what some may now call an act of vandalism, but it was in line with what was going on all over the country in the 1960s. Four chimney stacks remained standing but over the next month, one by one, they fell. Red House had held lots of memories for all of us; I am not sure of Father's deep-down feelings, but he said that he did not see any point in refurbishing it.

The empty site was ideally placed for a new house, which was built when brother Tom married in 1972. At one time he was thinking of a career in architecture so being involved in the design of the new Red House and overseeing its construction gave him a chance to express himself in a new home, which was certainly more comfortable and convenient than ever the old one could have been. The rubble from the old house was carted away and formed the foundations for new pig sheds at Jacksons.

As the pig enterprise grew, more pigs meant more pigmen and so a bungalow and two more pairs of cottages were built close by. We did much the same over at the Norton farm, the farmhouse coming down in 1969 to be replaced by a new four-bedroom, draught-proof house, with all mod-cons. We built two more pairs of houses in 1970, a pair at Wickham Skeith for pigmen and a pair at Earls Green, Bacton, where Paul Braybrook, involved in pig-shed repairs, moved in.

Typical pair of farm workers' cottages we built,
this pair built in 1968

Abbey Farmhouse, Wickham Skeith, 1950

Farm office with grinding shed behind. The grinding shed with
mill was demolished in 2017 and a new grainstore built
at the west end of farmyard

Up until the 1980s, when farms were sold, it was the
land that made the money, the farmhouse being included for
virtually nothing. Thus, having bought Rectory Farm in 1964
we came to own the old Bacton Rectory, a late Victorian
gothic-style building. It had been built in the days when the

rector would have had a coachman, a groom and horses and gardeners, and his wife a lady's maid. They would also have had a cook. It cost a pretty penny to maintain and keep warm as Reverend Bowden, on a much smaller stipend, found in the winters and in the particularly cold winter of 1947 when he and his wife huddled up in just a couple of rooms. When the church authorities sold it off with the land they built a new, smaller Rectory in the centre of the village much closer to the church. We eventually found a buyer for the old Rectory in 1970, a Hungarian, Mr Ludmani, who paid £5,000. He refurbished it and converted it into an old people's home, renaming it The Grange. Thirty years later, it was where Father spent his final days.

As we extended the pig operation further throughout the 1970s, we continued to build accommodation for the staff. By then we had come round to thinking that detached dwellings had attractions over the semi-detached houses which we had been building. We had had the occasional falling out between neighbours or more commonly between neighbours' wives: 'Good fences make good neighbours', is a saying which comes to mind. Several of our pig-farming friends had already come to that conclusion and, on journeys around the country, I was particularly impressed by some bungalows in Cheshire belonging to Philip Solari, a progressive pig farmer, who had a style of pig building named after him. He was not averse to us using his design for the two bungalows at Scholes, built in 1979, and for two more several years later at Mendlesham.

We have continued to make improvements to the existing housing stock. All our earlier houses had already been fitted with Raeburn stoves, which were popular with most tenants and provided hot water, cooking facilities and warmth to the kitchen. Unfortunately, many stoves were looking rather sad after two or three tenants had passed through, and electric cookers were the replacement of choice. Extra insulation has been added to the houses and the old metal Crittal windows replaced with smart-looking, double-glazed, draft-proof, PVC-framed windows. Very few of the houses were connected to the main sewer and we are in the process of replacing the old septic tanks with mechanical treatment plants.

In the early 2000s, my son William was responsible for overseeing the building of houses, including two on land in Wyverstone. One was subsequently sold and the other is now part of the portfolio of houses let out. We have fewer staff on the farm than several decades ago and thus a lesser need for housing for employees. However, there is a strong demand for rental accommodation locally and, although this was not our original intention, it has provided us with another income stream, most welcome in view of the forthcoming loss of the single farm payment following Brexit. This demand for rented residential accommodation has opened our eyes to other possibilities and we are actively pursuing the development of infill sites around the farms.

CHAPTER 8

CO-OPERATIVE ACTIVITIES

As a family business, we have been much involved in co-operative activities with our fellow farmers. Shortly after coming to Suffolk, my grandfather became one of the early members of Eastern Counties Farmers (ECF). This co-op started trading in 1904 and, until Fram Farmers was formed in 1960, we bought considerable quantities of animal feed and agricultural equipment from them. We were also very much involved with the co-operative bacon factory at Elmswell.

When my father, Tom Black, started keeping pigs he initially bought and sold them either through a local pig dealer or through one of the local livestock markets. In time he felt he would be better employed attending to matters at home rather than spending time seeing pigs sold in the market so decided to send them to the local bacon factory at Elmswell. Just before World War I, several pig farmers, dissatisfied with the prices they had been receiving for their pigs, had established this co-operative at Elmswell, buying pigs on a graded- and fixed-price basis. The local dealers, seeing that this would cut into their business, decried it, maintaining that they could pay a better price to the farmers for their pigs than the factory. Occasionally this would have been the case, for the factory was in a business which had its ups and downs, but, in most years, it was profitable and was able to pay a dividend. From Father's point of view, knowing that he could get a fixed price per pig was better than the uncertainty and occasional roguery of the open marketplace.

As time went by, Grandad became more and more interested in the pig processing business asking penetrating questions at the factory's open meetings, and in 1924 he was elected to the board which comprised participating farmers. Two years later he became chairman, a post which he held from 1926 to 1934 and again from 1941 to 1956. As chairman he became involved in the politics of pig production and processing both locally and nationally, so much so that he was appointed the first general manager of the Pigs Marketing Board, a national role which he held between 1934 and 1936. It was the time when government was looking at marketing boards to pull British agriculture out of the recession it had fallen into. My father was elected to the bacon factory board in 1933 and with the interest both he and Grandfather had in the factory, I became involved early on. I visited regularly and was appointed to the board in 1963. In this way, the family involvement was continuous from 1924 until the mid-1980s.

My first visit to the factory was in 1949 shortly after I left school. The heat, the noise and the smell, particularly in the slaughter hall, were nearly overpowering, but from there onward, having been a biology student, I found it fascinating to see the processes as the pigs were transformed into 'sides' ready to go into the curing tanks to come out as bacon four weeks later.

It was said at the factory that the only part of the pig not used was the squeal. The chief cuts were the back, belly, ham and the fore-end but mostly the grocer got a complete side delivered and it was up to him as to how it was cut up. Blood was dried and sold as a high-protein diet for pigs and chickens. Bones were ground up for fertiliser, the intestines were cleaned and used for sausage skins, while the kidneys, hearts and lungs also found their way into the food chain. The power for all the processes in the 1940s was provided by a diesel engine that was the pride and joy of the machinery department, who kept it meticulously cleaned and polished.

Pigs came in all shapes, sizes and colours with two or three from one farmer, four or five from another. It suited the farmer but was not the most economical way from the factory's point of view, which had to do the best it could from whatever was sent in. As a means of advertising what the factory was looking for, an annual show was introduced, but the coming of World War II put a stop to that, as to so many other things. However, once the War was over the show was reinstated with the same objective. Producers were invited to send in groups of four pigs, two of which were slaughtered, and their carcasses displayed, while the other two were shown live. Points were awarded and totted up to find a winner. The judges for the carcasses usually came from the Low Temperature Research Laboratory at Cambridge. They were involved because one of the objectives of the Laboratory's work was the preservation of food stuffs and, as bacon curing was part of their remit, they were often doing experimental work at Elmswell. At that time there were several abattoirs and processing plants locally and a manager from one of these abattoirs would usually be invited to judge the live pigs, for the type of pig acclaimed at many of the livestock shows was not necessarily the best suited for processing. There was a luncheon party held in a marquee for two or three hundred guests and Grandad would invite some notable agriculturalist to address the assembly. One year I was detailed to collect the cup we had won in one of the classes, presented by Mr Nugent, the Minister of Agriculture.

When I joined the board in 1963, Ken Crook had just been appointed general manager following the death of Brian Richardson. Bill Johnston (whose farm I had visited with the Young Farmers) was chairman. As a board member, I was in and out of the factory regularly and got on very well with Ken, but the role presented its challenges. Our farm was the largest supplier of pigs and at one time provided 40 per cent of the kill. Many of the other board members were relatively small producers and understandably wanted to get the highest possible price for their pigs all the time. This meant that it was difficult for the factory to put aside the funds necessary to invest in new buildings and machinery to keep abreast of its customers' requirements. The biggest customer at that time was M&S. The factory had moved on from selling sides of bacon into further processing and prepacking: sausages, pork pies, sliced bacon (both back and streaky) and cooked hams, which were particularly in demand for Christmas.

Theoretically, these were all higher-margin products. To produce them required new and often expensive machinery, so inevitably bank borrowings soared as indeed did the interest payable on the loans. Many areas needed attention, not least the necessity for pigs to be sold in balance. It was easier to sell backs than bellies, hams than fore-ends and one needed to be sure that high-value meat did not finish up in lower-value sausages. New equipment meant reorganisation, more staff and more training. Dealing with supermarkets and coping with their requirements was a very different business from supplying butchers' shops, of which there were becoming fewer and fewer.

Just before my time on the board, it had been decided to invest in a computer to assist in making these decisions and help with the invoicing. It was a big machine and had to go into a specially converted room with cooling facilities. It made for headache after headache. A hand-held computer nowadays would have been more easily programmed and would have much more computing power than that machine had. An enormous amount of management time was spent on it. It can be a difficult decision to know when it is the right time to embrace new technology and when to wait for further development to take place.

Profitability, as ever, was variable. Rebranding took place in 1971 to keep up with changes in the market, and was documented thus in the 1971 Annual Report and Accounts which read:

In June, St Edmunds changed their brand name to Farm Kitchen, and a completely new style for product packs was designed. The modernised label is well suited to marketing in the fast-moving convenience food sector in which the company now trades. Although the old St Edmunds label has become well established – having been used since the company's foundation 60 years ago – the change-over was well received by our customers and sales of our own branded products increased by 20 per cent.

This did not prevent the factory making a loss of £85,318 in 1972, a difficult year for all bacon factories in the country in spite of the introduction of a bacon stabiliser by the government in April 1971. A target price was introduced into the calculation and also mentioned in the report: *This target price was 2½ per cent below the average price of all top grades of bacon on the market each week and would be used instead of the UK price if the latter fell below the target price.*

My Grandfather, David Black, 1956

Me, 1960, receiving a trophy on behalf of David Black & Son Ltd
at the St. Edmundsbury Bacon Factory Show from
Sir Richard Nugent, Minister of Agriculture

A Black Sits on
The Board Again

THIRTY - YEAR - OLD MR. DAVID
BLACK, of Bacton, Suffolk—grandson of
the David Black so famous in the pig
industry, who died in 1960—has been
appointed to the board of the producer-
co-operative St. Edmunds Bacon Factory.
His appointment maintains the link which
the Black family has had with the factory
for 30 years.

*Mr. David Black. He carries on a 30-
year link.*

    The late David Black was more in-
terested in cows until his son TOM, a St.
Edmunds board member since 1934, began
pig keeping in 1922 and convinced his
father that pigs, sold on the hook through
the St. Edmunds factory, would make
more money. So the cows went, and since
the first delivery of Bacton pigs to St.
Edmunds in 1923, all baconers produced
have been sent to the same factory.

    It was in 1925 that the elder David
Black was elected to the St. Edmunds'
board. He was chairman from 1926 to
1934.

    Now the new board member, grandson
David, is in charge of the pigs at Bacton
and sends to St. Edmunds between 4,000
and 5,000 baconers a year—about one-
tenth of the factory's supply.

Article from the East Anglian Daily Times, 1962

Rebranding of St Edmundsbury Bacon Factory as
Farm Kitchen Foods in 1971

So it was that in 1974, looking around for other sources of finance, the board agreed a takeover by ECF, then under the management of Tom Thomas, effectively exchanging one farmers' co-op for another. Inevitably my involvement was much reduced, though we continued to be its largest supplier. I was disappointed that shortly after taking over, ECF dispensed with Ken Crook's services, replacing him with David Neal. ECF ran the factory well for 11 years investing in new equipment and extending buildings, and for a time it returned a healthy profit. However, by the mid-1980s, it, too, found the factory was a large drain on its capital resources and sold it to Hillsdown Holdings. David Neil was retained as the general manager and I remained involved, particularly through the work we were doing with M&S towards breeding a 'better' pig, but I was no longer on the board. Sometime afterwards, Hillsdown sold the factory to Grampian Holdings, but sadly, after only a short period, Grampian closed it and left it standing empty and unused. The site was eventually sold for housing development. We found new customers for our pigs in the shape of George Adams at Spalding and Bowes at Watton.

Our relationship with the Elmswell factory, like all close relationships, had had its ups and downs, but the assurance of a reasonable market with premiums for quality, virtually on our doorstep, was of great value to us over many years. Transport costs and weight losses were at a minimum, we were free enough to get on with the job of producing the pigs, and at the same time were close enough to have a feel for the market requirements of the factory and its customers.

In 1978 I was invited to join Terrington Experimental Husbandry Farm Pig Advisory Committee and continued there until the government decided to abandon the concept. There were several such farms over the country serving the useful purpose of putting into commercial practice the results of various research projects. They ran open days that gave farmers the opportunity of seeing and discussing which ideas showed promise and what we ought to be looking to do in the future. Terrington had experimental units for pigs and arable crops. On the pig side, among other activities, the farm conducted feeding trials, exhibitions of different housing and slurry systems, and held muck handling demonstrations.

We were also regular visitors to a similar experimental farm at Boxworth near Cambridge, which specialised in the evaluation of new varieties of wheat and barley and on the effect of different sprays and the timing of spraying on arable crops. Robert and I would also visit Stockbridge House in Yorkshire where we learnt a lot about rhubarb, though here the locals felt we were muscling in on their forced-rhubarb territory.

These experimental farms did much to bring new developments to the attention of farmers but by 2007 the government thought this work should be done and financed elsewhere and the farms were sold off. Fortunately, their role has been taken over by a range of ad hoc enterprises, such as Morley, near Attleborough, another farmers' co-operative operation, and by seed companies doing experimental work on individual farmer's farms.

In the 1960s we were still growing daffodils and tulips on the farm. Our customers were consolidating and, as the profit margin in these shrank, we joined Lingarden, a Spalding based co-operative. For several years we sold our bulbs through them until we decided that the relatively heavy soil at Bacton was not the best for this operation and abandoned these flower crops altogether.

For a while we grew special stock strawberry plants. Situated out of the main strawberry growing area, we were able to multiply virus-free stock without getting the plants infected. This necessitated us joining the Nuclear Stock Association, an organisation based at the National Institute of Agricultural Botany (NIAB) in Cambridge. The institute supplied the virus-free stock produced from meristematic[1]

---

[1] Meristematic stem culture: meristematic cells are produced at the very tips of roots or shoots. Even if the plant is infected with a virus, these tips

stem culture. We multiplied the stock which was inspected twice during the growing season by inspectors from NIAB. If any plants were suspected of showing any sign of disease they had to be taken out and destroyed. We lifted the runners in October and November, bunched the plants into bundles of 25, then sold and delivered them to strawberry growers in the Wisbech area, 1,500 plants here, 5,000 there. Many of our customers were small-scale growers picking fruit for Wisbech's jam factories. In one year, we sold 750,000 strawberry plants. Slowly, however, the jam trade fell away, as did the growers, and we eventually made the decision to pack up that business.

It is often said that it takes an outsider to initiate change and it was Stephen Horvat who was the instigator of Fram Farmers, a co-operative group of farmers he brought together in 1960. A year or so after its formation I joined the board, which comprised an interesting group of farmers, and I learnt much from them. Initially they were Stephen as chairman, Ivor Bull, a former tea planter, as vice-chairman, and, among the other influential people involved, Reg Harvey, Alec Cooper, George Rous, Anthony Daniell and Bill Green. Anthony had been a successful submarine captain in the 10th Flotilla during the War and was awarded the DSO when invalided out of the Navy before starting to farm in Suffolk.

Stephen Horvat was very much the driving force of Fram Farmers. One of his many sayings was, '*Today is late enough; do it now,*' and in time this was adopted as our motto. He managed to persuade 19 local farmers to meet at the Crown and Anchor Hotel in Framlingham on 17 March 1960. We agreed to form a co-operative society, making use of grant aid which was available, with the aim of reducing farmers' costs by bulk-buying items such as sprays, fertilisers, fuels and feeding stuffs. A committee was formed with Dick Leach, then the NFU secretary for the Framlingham area, as

---

are free from the infection and therefore if used for propagation will produce virus free stock

secretary. A £5 subscription started the funding. Fram Farmers was to be run on a shoestring with much unpaid support from the initial subscribers. Various local suppliers were contacted and offered larger volumes of business and opportunities to cut their costs if they dealt with this new group. They would receive orders in advance, deliver full lorry loads, not be expected to offer after-sales service and be given cash on the nail, all in exchange for keener prices for the farmers. Enough suppliers responded positively that the group was able to offer its members lower prices for fuel, fertilisers, baler twine, tyres and spray chemicals, and in due course other commodities.

Inevitably there was controversy. Some members of ECF, a longer-established farmers co-op, felt that this new venture was muscling in on their territory and would cherry pick, potentially undermining their business. As a result, they would have nothing to do with it. Sadly, ECF, in common with many others in the supply chain, had allowed their overheads to build up and they had become much like any other merchanting business. Salesmen called on farmers to get their orders, there were large stocks of machinery, some which didn't turn over very quickly, there were second-hand machines which had been taken in part-exchange, often on the books at a higher value than that for which they would ultimately be sold, and although this was convenient to ECF members, it came at a cost.

It was not easy for Fram Farmers in the first few years, with some suppliers being downright hostile, and it seemed that they would have been pleased to see the new venture strangled at birth. Some sceptics felt that it was only Stephen Horvat wanting to feather his own nest. However, slowly but surely business increased. A full-time manager, Peter Lusted, was employed and, although he knew nothing of agriculture, he was willing to learn and took on board the 'lean and mean' philosophy that Stephen was impressing on the new buying group.

Initially meetings were held in Stephen's home in Worlingworth, then in one of Reg Harvey's houses before Peter Lusted hosted the office in his bungalow for a while. Stephen was determined to put Fram Farmers on a more professional footing and so when 4 Market Hill in Framlingham came up for sale in 1965 he went out and agreed to buy it without consulting the board. I suggested to Stephen that this may cause some difficulty for him at the next meeting, having heard chatter that some thought it would cost far too much to convert to offices. In true Stephen style he was able to persuade the local builder, Mr Gobbit, to prepare a quotation almost immediately for the conversion work required and presented the whole package for approval at the next meeting. Although Stephen came in for much criticism at that meeting he got his way as usual. He said that if the board would not sanction the deal he would buy the property and rent it to the group. It worked out very well, particularly as Fram Farmers was able to rent out some of the space in the property, which paid the interest on the money borrowed. Eventually, as the business grew and more staff were employed, the sale of 4 Market Hill helped finance the purchase of larger premises at the Old Laundry in New Road, Framlingham, which housed the offices from 1977.

Due largely to the efforts of Stephen and Peter Lusted the membership increased along with the turnover. At the end of December 1960 after just seven months' trading, Fram Farmers had a turnover of £64,000, the membership had risen to 89 and the cost of running the business was 1.49 per cent of turnover, a figure that Stephen was intent on not exceeding. By 1965 the turnover had gone up to £780,000 and membership to 250, and by 1970 it was £1,728,000 and 308 farmers. The cost of running it had come down to 1.18 per cent of turnover.

A little later Stephen Wise, a leading farmer from West Suffolk, came in as a director. He and Basil Ambrose, also from Cavendish, had started the Agricultural Spraying Company, which sold spray chemicals and offered a spraying service – to members of Fram Farmers to the benefit of both parties and to others who they encouraged to join the co-op. Stephen Wise's parents were refugees from Czechoslovakia who had moved to England and started farming in Cavendish in the middle of the 1930s. Like Stephen Horvat he was always on the lookout for new ideas and travelled frequently to the States from where he brought back many bright ideas, and a wife! Indeed, from my experiences in the States on the Young Farmers' exchange I too found that practices there often turned up in the UK some 10 years later, having been brought back by entrepreneurs such as Stephen.

On the back of the buying group, other opportunities for co-operative activities within Fram Farmers arose. Porcofram was one of the first. Several pig farmers, although not willing to commit themselves to the co-operative bacon factory at Elmswell, set up a group to market their pigs on a weekly basis to other pig slaughterers. They had their teething troubles with the first manager who spent more time off the job than on it. However, after he left the business developed and prospered. I did not get involved with Porcofram as by then I was on the board of the bacon factory at Elmswell so there would have been a conflict of interest.

Another group was set up to grow and market vegetables but this was not very successful, as harvesting potatoes, carrots and maize in a wet autumn on heavy land proved challenging. However, this morphed into a pea-vining and rape-swathing syndicate. Grain handling, seed dressing and an agronomy group were examples of other activities investigated and developed within the co-operative, while a cash and carry was instigated at 4 Market Hill.

One area I was involved with was 'earth wearing' parts. Plough shares, breasts and cultivator points on farm machinery were always wearing out and in need of replacing. Most of the cultivation equipment was made by Ransomes, Simms and Jeffries of Ipswich which had a virtual monopoly on the supply of spare parts. George Overton, who had experience in the Sheffield steel industry, offered his services to Fram Farmers and introduced us to some of the Sheffield firms who were looking to diversify into casting agricultural replacement parts. Obviously we as farmers had queries as to whether the quality was as good as the parts supplied by Ransomes. Earle Morley and I visited and, being sufficiently impressed, offered our farms as guinea pigs to try out the replacement parts from a couple of the Sheffield firms. These proved to last for as long as the original Ransomes parts but at a fraction of the cost and thus was born another line for Fram Farmers: Mercofram. I was chairman of this group and over the following years we built up a big turnover for Fram Farmers by sourcing these and many other parts from Sheffield.

Other co-operatives had formed or were forming in Suffolk. Representatives from one or two came to speak to us at Framlingham, wondering if there might be benefits in some form of working together. Egos often had to be soothed but we did deals with some, including Halesworth Farmers, Blyfarmers and a group in the Sudbury area. We also had talks with Loddon Farmers and Mid-Norfolk Farmers, but we didn't quite see eye to eye and eventually in 2003 they joined other Norfolk groups to form Anglia Farmers, based just outside Norwich.

Prior to Blyfarmers joining Fram Farmers, the two co-ops had worked together in a joint venture, Blycorn. I was not much involved in this because we used most of our own grain to feed to our pigs and consequently had little grain to sell. Some farmers, however, felt themselves at the mercy of grain merchants and traders, came together and, by offering larger quantities and guarantees through Blycorn, managed to get a better price in the market.

It was a busy time, for apart from helping Peter Lusted sort out the relationships between these diverse activities, the board became involved with recruiting extra staff and sorting out finance, determined that costs did not get out of hand. We kept overheads to a minimum and forged close links with the banks, which provided us with some working capital. In addition, we all had our own businesses to run.

The First Board, Fram Farmers 1960-1967
Standing L-R: M.A. Lloyds (Secretary), P. Lusted
(General Manager), D. Black, Cdr. J.C. Jacob, K.W. Bullock,
J.H. Mayhew, S.G. Wise, J. Kerr, H.W. Kindred, J.N. Holmes,
T. Denham-Smith, T.L.V. Brown, Maj. G.N. Rous
Sitting L-R: R.R. Harvey, A.K. Cooper, Cdr. A.R. Daniell
(Chairman), S.N.Horvat (President), I. Bull, R. Leach,
C.W.C. Green (Vice-Chairman)

In the early 1970s Fram Farmers underwent a significant restructuring, catalysed by the death of Stephen Horvat, aged only 60, and by two other senior figures, Peter Lusted and George Overton, deciding that they wished to move on. Stephen's death came as a shock to us all, but apart from living at a frenetic rate, he smoked like a chimney, so maybe we should not have been so surprised. Much of that future development of Fram Farmers is detailed in a short book, *A Story of Opposites*, which tells Fram Farmers' story from 1960 to 2000 as seen from a range of viewpoints. Suffice to say that we appointed David Thirkell as the new general manager in 1971. He embraced the co-operative vision and worked very effectively with the board for the next 20 years.

David recalls in his chapter[2] in that book that there were 'communication' problems when he first came to Suffolk. He and his wife Glenda took the opportunity to meet members including me at the 1971 Suffolk Show. He wrote:

Glenda and I met David Black for the first time. Glenda comes of farming stock, so she asked him what his main enterprises were. "Pigs", says he. "Oh, do you have many?" says she. "Several", came the reply. Now to us, the word several meant more than one but not many more. Need I say more? We went for some time thinking that the Blacks had a few pigs in the back yard.

Over the years Fram Farmers has become sufficiently well known in the trade and, by guaranteeing a big enough volume of business, has been able to negotiate annually an overriding discount which has been returned to members retrospectively. The scale of our own business dealings with Fram Farmers has been such that in the third year of Fram Farmers' existence, we were doing about £300,000 turnover with the group, mainly barley for feeding pigs, and by 2013 it was £2.5 million. Many of our requisites were bought through the group and for many years we were its biggest customer. However, following the change in our milling arrangements on the farm in 2017, our turnover with Fram Farmers for 2020 was down to £1.14 million.

We started growing oil seed rape in 1993 and joined United Oilseeds, a co-operative set up specifically for the marketing of the crop; we are still selling through them, 30 years later. There have been interesting genetic improvements in this crop over the years. It is divided into two separate types, one with high erucic acid for industrial

[2] Thirkell, D. 'The Expansion Period' in J. Ball (Ed) 2000, *A story of Opposites: Framlingham Farmers 1960–2000.*

oils and the other with low erucic acid suitable for producing oil for human consumption. Although the extracted residue in both cases goes for animal feed it does not do to get the two sorts mixed. As of 2020 the future of the crop is not assured because it is very susceptible to flea beetle damage which can result in total crop failure. The neonicotinoids with which the seed had been dressed and that we had been using to control the flea beetle have now been banned as it appears that they may have been causing harm to bees. We are now looking for alternative control measures.

When Birds Eye set up a freezing plant for peas in the Lowestoft area in 1949 and simultaneously offered contracts to local farmers to grow peas for them we were envious of those farmers able to grow the crop. At the time we were looking for a profitable break crop as part of an arable rotation. Sugar beet harvested with a single-row machine could cause a lot of soil damage in a wet autumn and was not particularly profitable. Peas appealed, being a profitable leguminous crop putting nitrogen back into the soil. Peas and beans were spring crops, harvested in June and early July, which meant that muck could be spread on the stubble and the land ploughed before the rush of corn harvest, providing a valuable entry for drilling winter wheat. This also relieved the autumn workload and provided an opportunity to reduce the population of such weeds as black grass by allowing their seeds to germinate and then killing them off before drilling the next crop, which was usually wheat.

As we were on the west side of the A140 we were just outside Birds Eye's catchment area and it was not until some years later, when other firms involved in freezing stepped into the market and offered contracts, that we were able to begin growing peas. It was a crop that needed a fair bit of organisation since the drilling had to be done in such a way as to produce a succession of crops of the right maturity throughout the harvesting season. Once started, harvesting continued 24 hours a day, seven days a week, moving from field to field based on the tenderness of the peas, measured by a tenderometer. Left even half a day too long the quality could quickly deteriorate. Initially the crop was cut and carted to a central vining station, but by the time we were involved in the 1970s, mobile machines had come onto the scene, reducing much of the hand work. The peas had to be delivered to the freezing factory within a couple of hours of being shelled, so it made sense to have two or three viners working simultaneously in the same field to keep the transport costs down. In turn this meant that a large acreage of peas and many individual farmers were needed.

I was keen for us to grow peas and worked with neighbouring farmers to set up a local pea group. We were encouraged by the Central Council for Agricultural Co-operation as represented by Ken Leggett and others and visited several independent pea-growing groups including Pembertons of Cambridge to seek their experiences and learn what we could. A group of 10 local farmers who were together prepared to commit 750 acres to the project formed the group and signed a joint and several guarantee. The original pea group included us, Edmund Baker, Keith Bullock, John Forrest, John Gooderham, William Hamilton (Mary's brother), Ivan Lockett, Nigel Merriam, Ron Nunn and Len Waspe. As can be imagined, there was much work to be done. Farmers' meetings lasted well into the night while we were establishing the group; we organised finance and negotiated contracts, set up the plant and recruited staff. With everyone prepared to do their bit, it got off the ground and as a relatively small group we had a harmonious relationship. The group bought some second-hand machinery from Louth in Lincolnshire and we, represented by Les Gooderham, our workshop foreman, were much involved in installing the plant at Keith Bullock's Boundary Farm in Cotton. Here at 'the chiller' we were able to wash the peas from the mobile viners, cool them and then deliver them to the freezing plants at Lowestoft or Kings Lynn within the tight time frame demanded to ensure freshness and tenderness. Keith took responsibility for the chiller and its development. The site of the chiller on his farm suited us well, as he and his family were milk producers and any wastewater from the plant could be irrigated onto the grass meadows for the cows and any waste peas and haulm went directly to supplement the cows' feed.

Gipping Valley Growers (GVG), for so we came to be called, sold to a variety of processors, both canners and freezers and it was Ronnie Nunn and John Forrest, and then Nigel Merriam, who negotiated the contracts with them. The freezers were looking for a younger, sweeter pea, compared with the canners, so it was helpful to have a contract with the canners for those peas with a higher tenderometer reading in case harvest could not take place at peak time for the freezers. Often GVG had four contracts to meet. Initially the main customers were Anglia Canners at Wisbech, Mortons (Beechams) and the Co-operative Wholesale Society. When we started, the can cost 6p and we were paid 5p for the peas inside. It seemed a good idea at the time when, in 1981, we began selling peas to the public from the chiller. It started with a flourish but became more of a diversion than of value so was stopped after only a few years.

The pea harvest ran for 24 hours a day throughout the vining season, from June to the end of July, so any breakdowns, either in the field or at the chiller, needed fixing right away. Our workshop team fitted out the mobile workshop which accompanied the cutters and viners wherever they were working. John Gooderham's mechanic, George Freeman, was the night foreman and Bob Willis, who was employed independently by GVG, was the day foreman and chief mechanic. Initially the peas were swathed with cutters and then vined with trailed harvesters. In 1979 the group bought its first 'podder', a self-propelled harvester with a much greater output. In 1976 GVG was further enlarged, with messers Balls, Davey, Fenning, Ibbot, Miles, Pulham and Reeves joining. Some of the members, including us, also grew beans for GVG under contract for freezing, but they were not under quite the same time restraints as the peas.

Pea Viner used by Gipping Valley Growers, 2000

At the height of production in 1984 the group produced 2,200 tons of peas and 1,200 tons of beans. Farmers from within the group supplied most of the labour. I had spent a great deal of time getting the peas up and running but stepped back as brother Tom, and then nephew James on the financial side, became more involved. As time went on we gained experience and doubled the acreage, and the operation ran successfully for about 30 years. There were exceptional, profitable years in both 1984 and 1996 but in both cases punitive cuts in tonnage and prices followed. In early 1997 Anglia Processors went into receivership owing £122,359. Despite this the 1996 crop had given growers the best return ever. However, the demand for peas was falling. The supermarkets were increasing their sales of fresh vegetables which they were able to supply all the year round, importing where necessary to maintain continuity. Consolidation followed and many of the canneries were taken over by Hillsdown Holdings which later sold them to Premier Foods.

At a meeting in 2000 chaired by Nigel Merriam, our hard-working chairman of 13 years, GVG reported the shrinking market. The price obtainable dropped considerably and, by 2002–2003 GVG and the farmers growing the peas were making losses. A new, self-propelled pea harvester would have cost in excess of £250,000 and with no future prospect for higher prices and in view of uncertain future returns the decision was taken to wind up the group. It was a sad day but it had become uneconomic to continue.

We still have dealings with United Oilseeds but other than that, of all the co-ops that I have been closely involved with, only Fram Farmers remains. It has adapted and changed over the years and continues to provide a great service to its members, 89 in 1960 and over 1,200 in 2020. Considering how many similar organisations have fallen by the wayside, it is a testimony to the foresight and perseverance of those who have given their time and energy to it over the last 60 years.

CHAPTER 9

FAMILY LIFE

Mary and I celebrated our Diamond Wedding Anniversary in 2017 but despite the passage of more than 60 years I clearly remember the early days of our relationship. We had gone out together a bit before my trip to the States in 1954 and we wrote to each other while I was away. Soon after my return we arranged to meet again. Mary was studying at the School of Domestic Economy at Eastbourne and, as I would be travelling from Suffolk, we planned a day together in London. We saw Dame Sybil Thorndike in *A Day by the Sea* at the Haymarket Theatre and ate out at the Trocadero. That was a rather more romantic date than the time I took her to Slough when I had to pick up some parts for the grass-drying plant, or when I asked her to come with me in the farm's green van to take four pigs to the pig-testing station at Selby. Mary may have been a farmer's daughter but she recalls being rather embarrassed smelling of pig when we stopped at the Station Hotel in York for a meal. Not all our courting was farm based, for over those few years we also attended numerous parties and dances together. After one party at the Leiper family's home in Weeley, we swam in the sea at Clacton. The moon was shining, the water sparkled; it was magical.

Mary had moved with her family from Cotton Hall to the Rosery, Little Stonham, in 1954 when her father took on a larger acreage farm. It was at the Rosery that I proposed to Mary one evening in September 1956. I was 24 and she was 19. Once she had said 'yes' we woke up her parents to share our good news and to get their blessing. I was often at the

Rosery and one evening, showing off as occasionally I was prone to do, I did a handstand in the middle of the drawing room. Unfortunately, it was immediately under the crystal light fitting with catastrophic results, leaving a very deflated David. I apologised most profusely to her parents. They were very forgiving, and we always got on very well.

Mary and I were married by the Reverend Frost at what was then the Congregational Church in Stowmarket on 25 September 1957. Mary's father, like my grandfather, was a Scotsman and had been brought up in the Presbyterian church of Scotland. The nearest Presbyterian church was in Ipswich which they thought was rather too far to go each Sunday from Cotton and subsequently from Little Stonham, and so they had chosen to worship in Stowmarket, finding the Congregational Church a fair substitute for the Presbyterian style they were used to. My paternal great-grandfather had from time to time preached in his local Presbyterian church in Scotland and knew his Bible well, as did my father. However, it was my mother who was the regular attender at St Mary's in Bacton, and who brought us up to be members of the Church of England. She suggested to Mary and Miriam, Robert's fiancée, that they should consider being confirmed into the C of E, so that we could all worship together at St Mary's, and so it was that the Reverend Gordon Budd gave them preparation classes together and they were confirmed in St Mary's the year after we married. Our wedding day had dawned wet but that did not spoil the occasion. The bridal party briefly stopped the traffic in the middle of the town, the main route from Ipswich to Bury. My brother Robert was best man, and Mary's bridesmaids were her cousins Nancy Hill and Jill Sinclair, together with a very young Helen Leiper, daughter of Mary's cousin Ena.

Once the wedding ceremony was over, we exited the church through an arch of hockey sticks, courtesy of my friends from Stowmarket Hockey Club, before going on to the reception at the Pier Pavilion in Felixstowe to celebrate.

Eventually, Mary and I drove away, borrowing Mother's Hillman Minx, to Ingatestone, where we had booked to spend the first night of our honeymoon. Ever practical, a few weeks before our wedding when Mary and I had been returning from a day out in London, we passed a hotel called The Chase in Ingatestone. I pulled up at the door and suggested she book us a room for our wedding night. The receptionist had offered only two rooms for that date and Mary chose the least expensive one. When we arrived on our wedding night, we were somewhat disappointed to find that we had to go through another bedroom to reach ours, but fortunately it was unoccupied. I remember that we were rather embarrassed by the large amounts of confetti that fell from our clothes and spent some time clearing it up before heading down for breakfast the following morning.

Ever since I had gone away to school Mother had packed my suitcase and she had done so again for my honeymoon. We only discovered the next morning that there was no tie. It was not the done thing to go tieless in those days and so Mary and I had to go shopping in Brentwood before driving on to Tilbury to take a ferry across the Thames – no bridge or tunnel then. From there we continued to Dover for the crossing to Calais before motoring on to Paris. I should add that once I was married, Mary always did the packing for our holidays, with me usually on the farm until the last minute when I would rush home to load the car. Many years later, given a choice of two, I put in the wrong suitcase. We arrived in Ludlow before discovering the mistake. None of our night clothes were there nor any of the other presumed essentials. I was sent out to garner what I could in the 15 minutes before closing time; not surprisingly that was a particularly frosty night!

To return to Paris. We spent three days there, though this turned out not to be the romantic experience we hoped. A dodgy tummy is no fun at any time and unfortunately Mary suffered from eating some questionable pâté at Versailles on the first day. Thankfully she was recovering well when we motored on to Nice, leaving the sights of Paris for another day. As we drove south we passed a car with UK number plates, then a little later it passed us. This happened two or three times until we stopped at the same petrol filling station. It turned out that the occupants were also a honeymoon couple, from Ayr, who were on their way to Nice, too. We arranged to meet up in Nice; they were taking the main roads and we were going by way of Vichy, La Chaise-Dieu to view wall paintings, Le Pont du Gard to see the Roman aqueduct, and Avignon, with its famous bridge (as in the song) and Le Palais des Papes. Several years later, on one of our visits to Scotland, we caught up with the couple in Ayr. Over the years we have had a surprising number of chance encounters such as this which have always added extra interest to our travels.

After a week of swimming, lazing and getting to know each other better, we returned through Grasse, the world's perfume capital, where we stopped to see white jasmine flowers being picked for distilling. Then, zigzagging through mountains, we followed the route which Napoleon took after his escape from Elba as far as Grenoble, then drove north through the rest of France before returning home for our first night in our new house, which has been our home ever since. Both sets of parents were there to greet us. Mary's mother had made up our bed and together they had prepared a meal which we all shared, the first of many in our own dining room.

We settled in and over the next five years produced four children, Elizabeth, Christine, Alison and William. My parents rarely drank alcohol but when William was born they had a celebratory sherry. Both sets of parents were an enormous help in our early married life, babysitting, cooking, mending, entertaining.

The children remember many meals at the Rosery, Mary's mother being a proficient and generous cook. In winter, high tea would be a main course and it would be either fish pie or a dish she called 'sausage surprise'; Father-in-Law would playfully say 'the surprise was finding the sausage' but it was a favourite with us all. This would be followed by a wide selection of cakes including melt-in-your-mouth meringues, pancakes with homemade strawberry jam, and toffee biscuit finger, with homemade lemon to drink. In summer she would make a meat loaf to serve with salad followed by a similar spread of delicious cakes. After tea, Father-in-Law and I would look around his farm and discuss farming matters while the children played games or explored the outhouses in the farmyard. The children became used to my detours as I would often stop to look at some of our crops on the way back home. One dark winter night we got stuck in a drift of snow in Cotton Lane and had to be pulled out with John Davey's tractor. We used to have much more snow then and for many years barriers were erected on the Rosery land at Little Stonham to prevent snow drifting over and blocking the road to Norwich.

Robert and Miriam had four boys, James, Andrew, David and Peter. They were similar in age to our four children and they all spent much time together playing around the farm. The mind boggles at the thought of a gang of eight children running and cycling about freely on a farm today but then it was the natural thing to do. There was always something of interest going on and there were wooden boxes, bricks and straw bales for building dens, and when they were thirsty their Granny Black, my mother, was always happy to see them at Somerville, supplying them with 'fizzy beer' (orange squash with a dash of Eno's) and a scone. Before they left on their bikes she would offer them a dip in the sweetie bowl, always insisting they 'take another one to keep you going until lunchtime'.

When the time came for formal education, the children first attended Joy Mulley's nursery school, conveniently situated in the village. Joy was the daughter of the local carpenter, decorator and village undertaker, and the nursery school she ran was well-regarded. Consecutively our four went there along with their cousins. Joy recommended that they progress to Hillcroft, a prep school run by the Reeders in Stowmarket. They did well in both places and had a very good grounding. Both Mary and I had been away at boarding school and thought it had something to offer our children not available at the local grammar, so as Elizbeth approached her 11th birthday we looked at a few possible schools. Felixstowe College and St Felix at Southwold were recommended, and we visited both, taking Elizabeth with us. At St Felix, the headmistress, after a brief interview, took us round the premises but told Elizabeth she was to remain in another room until we returned. We felt that it would have been helpful for her to have seen the facilities, too, and we left feeling that it was cold and not very friendly. Felixstowe was more to our liking but still did not feel quite right for us. When talking with Stephen Horvat, he suggested we look at Runton Hill on the North Norfolk coast where his daughter, Wanda, was a pupil. 'It has a much more intimate feel. You will like the headmistress and the girls are confident,' he said. We visited and were impressed with the atmosphere. Elizabeth said she preferred Runton Hill so she and in due course Christine and Alison continued their schooling there, while William followed me, my brothers and his cousins to Culford, starting in Cadogan House aged eight.

While the children were young we put a tennis court in the garden at Harpsfield and later a swimming pool. These were both a great boon, especially as the summer holidays coincided with harvest time when I was busier than usual on the farm. Before that we often used to go to swim in the pool at Culford School on a Saturday afternoon, meeting up with other families there. My brother, Thomas, 14 years my junior would sometimes come too, providing an extra pair of hands and adding to the fun of the outing. As a teenager he had become diabetic and from then on needed to inject himself with insulin twice a day and ideally have a regular routine. As a result, when Mother and Father went away on farm visits or on holiday, Tom would come to stay with us at Harpsfield. He was good company and willing to play card games and such like with the children. When he left school, Thomas was considering an architectural career but it was felt that he would be safer working with family as we were all aware that he could have a hypoglycaemic condition when his blood sugar was low, which could put him in a coma. So, when he joined the farm, many of the vehicles were equipped with a tin of boiled sweets and with that precaution he got on with it. He went through a somewhat similar introduction to farming as we two older brothers had experienced, as dogsbody to whatever needed doing. He too became an enthusiastic member of Stowmarket Young Farmers and was county chairman in the early 1970s. Farming-wise he had an opportunity to take a lead when he took over the responsibility for the peas and beans for us, and became fieldsman for the whole group, making decisions about when to sow and when to harvest in co-operation with all the member farmers.

Thomas married Rosemary Ince in 1972. They had met through Young Farmers where they were both much involved in the public speaking competitions. Our three girls were bridesmaids at their wedding. Rosy had spent time in Papua New Guinea with Voluntary Service Overseas, teaching in bush schools linked to an Anglican mission station, and then taught at Bacton Primary School until they married. Their first son, Mark, was born in 1973, followed by twins Katherine and John in 1975. As well as his involvement with the pea group, Tom was responsible for the introduction of tramlines into our arable cropping, a system which is now almost universal, and he planted a stand of trees along the Walk, where the Avenue had been when I was a child.

Mary and I made time to go away on holiday with the children, usually at Whitsun. Winterton on the Norfolk coast was an early destination and taking the chain ferry over the river Yare at Reedham greatly added to the interest of the journey. Once at the coast we would explore the dunes, play on the sand and swim in the sea. In the evening it was possible to put the children to bed and for Mary and me to have a relaxing meal in the dining room as we watched the ships go by through the picture windows. One year we flew to Jersey from Ipswich airport, taking off from the grass runway which now, sadly, has been built on. That year, William, aged one, was considered too young to take if we were to have a relaxing holiday so he spent the week with Mary's mother, Granny Ham as the children affectionately called her, while we and the three girls went swimming and fishing for shrimps among the rocks at La Corbiere.

Another year, we collected the children from Hillcroft School on the Friday at the beginning of the Whitsun half term and drove by way of Stonehenge. We then spent some time with Mary's schoolfriend Rachel Willcox and her husband David, a farm machinery dealer near Taunton, before continuing onto Paignton for a week's holiday. We visited Brixham and the model village at Babbacombe and played with buckets and spades on the beaches. We bought tickets for a Val Doonican concert at Torquay's Princess Theatre, for the Saturday evening at the end of the holiday. Mary arranged to keep one bedroom in the hotel for the afternoon so we could all have a rest, making it possible for the children to stay up late and for us to then drive home overnight at the end of the concert. We had packed the footwells in the back of the car with luggage so the children could lie down for the journey. They must have slept well; on arriving home we lifted them into their beds before Mary and I also fell sound asleep, but within a couple of hours the children were wide awake and keen to go to Sunday School prizegiving.

Our Wedding, 25 September 1957

Children at the Rosery, 1963, after high tea, bathed ready for
journey home and bed. L-R: Alison, Elizabeth, William, Christine

Me with our three daughters at the Stowmarket Flower
and Produce Show, 1963

Flying to Jersey from Ipswich Airport, 1964

In Jersey, 1964, with Elizabeth, Christine and
Alison, and Jersey cows

The children at Stonehenge in their Hillcroft School uniform, 1967

Fun in the garden at Somerville, 1966 (back row: Mother and Mary, front row: L-R: Elizabeth, James, Christine, Andrew, Alison, I.David, William, Peter, Miriam)

The children, on holiday in Brixham, 1968

We also enjoyed two family holidays on the Thames, hiring a motorised boat called *Maid Fortuna* one year and another called *Maid Fidelia* the next. On the first, we started from Thames Ditton and explored a history of England, visiting Runnymede, Hampton Court Palace, Eton and Windsor. The following year we boated from Abingdon to Oxford and on to the upper reaches of the river as far as Lechlade. I remember an enjoyable afternoon spent with Father's cousin, Peter Church[1], his wife Josie and son Andrew, swimming in the pool at St Edward's school, Oxford, where Peter was a housemaster. One evening, moored for the night, Mary took Elizabeth with her to hunt out supper while I was left in charge of the younger three. Inevitably I was reading a book while they played happily outside, jumping to and from the boat to the tow path. Luckily, when William, only five at the time, misjudged the distance he was spared a wet landing by a quick-thinking, bowler-hatted commuter walking along the tow path. He simply extended his umbrella, handle first, and caught William by the strap on his buoyancy aid. I don't think this story loses anything in the telling, any more than another which does the rounds at my expense. On another occasion I was also supposedly keeping an eye on the children, this time on a post-harvest August bank holiday camping trip to Dunwich, when William had another narrow escape. We were walking along the edge of the cliff when suddenly it collapsed beneath him. Fortunately, it was a low, sandy cliff and he only tumbled a few feet before landing on soft sand with no damage done. That Dunwich camping weekend was a trial run prior to a camping holiday along the Loire in France the following summer.

We took the hovercraft over the Channel, a new experience for all of us, and drove to the Loire. One of the highlights of the holiday for me was a spectacular evening in the grounds

---

[1] Peter Church was the son of my Aunt, Mary Church, a half-sister of my grandfather, and Uncle George Church, who had been headmaster at Bacton School when I was a pupil.

of the chateau d'Azay le Rideau for *Son et Lumiere.* The chateau is one of the smaller ones, but is an absolute gem. The commentary was in French; we missed the finer points but grasped the broad outline of its history. A day or two later we called on Monsieur and Madame Delbos who lived near Cognac. They were farmer friends and customers of Bernard Barrabe. Bernard, whom we knew well, worked for the feed company Sanders, which was based in Rouen. Mme Delbos suggested that the children might like to camp in the garden for the night, offering Mary and me a bed in the house. Mary was thrilled to have a night off the ground and accepted with enthusiasm. I pitched the tent with the children while Mary was taken indoors to choose a bedroom. She was shown a choice of bedrooms and was taken aback by the woodworm evident in the bedsteads in a little-used wing of the house but eventually came across one with a brass bedstead. The cobwebs, dust sheets and black widow's dresses overflowing the wardrobe were somewhat off putting, reminiscent of Miss Havisham's house in *Great Expectations*, but it was too late to decline the offer of sleeping in the house and so she settled for the room with the brass bedstead. Meanwhile, out in the garden, the children discovered ants swarming all over the tent with the groundsheet a dark mass of them. We had managed to pitch the tent over a large ants' nest but fortunately there was no undue panic and we moved the tent before heading out for the evening. M and Mme Delbos took us to a firework festival nearby. It was a tremendous display attended by thousands. A fun fair was part of the festivities and the children were amazed as Mme Delbos bought a bag full of tokens for the dodgems, encouraging them to have as many goes as possible rather than the one or two rides they would usually have been permitted. As we left at the end of the evening the crush through the narrow exit was frightening. I remember having William on my shoulders and Mary and I holding onto the girls' hands for dear life. We were carried along by the crowd but I feared that if anyone had fallen over they would have been badly trampled.

After that holiday Mary said 'no' to more camping. Out for dinner one evening at Peter and Peggy Beaney's soon afterwards, Peggy mentioned that they were selling their caravan. The next day Mary went to look at it and came home triumphant. This gave us a base for a trip to the Lake District, at Keswick near Derwent Water early one summer. We also took it to a site in London for a couple of nights while we had a German girl, Elena, staying with us for an exchange visit with Elizabeth. We took her to see the usual sights and to the Royal Tournament held at Earls Court, a great military display which featured competitions between various branches of the armed services. During one breakfast in the caravan Elena spread marmite thickly on her toast thinking it was treacle; after recovering from the shock of her first bite, she took it in good humour, able to laugh at her mistake.

It was in April 1972 that we splashed out on a cruise on the Reina del Mar, embarking at Southampton and heading out to the Atlantic. We'd had an early start from home and the children were all very excited. Mary had spent weeks getting the clothes sorted and the packing done, so it was with some dismay that, when I reached for my wallet in my inside jacket pocket at the check-in desk, five pairs of eager eyes on me, I could not find the tickets. I felt again in that pocket, emptied out my wallet, searched every pocket, but they were not there. Mary and I had a brief exchange, wondering where they had last been seen. The study desk was the conclusion and it was all my fault for not having picked them up. The children had memorised the cabin numbers and recited these to the Union Castle Line staff at the desk, who were able to look on the manifest, check against our passports and allow us to board. The family behind us in the queue, the Shirleys from Kent, also with four children, made a joke of it and were most understanding. I remember years later calling on them at their garden nursery in Kent.

In Lanzarote we rode on camels up to the Fire Mountains. We were impressed by the volcanic activity and amazed to see our guide pour water through a shaft in the ground onto the magma, sending up a geyser. The children were fascinated to see meat being barbecued over the earth's heat there too. Casablanca in Morocco was another port of call and here we bargained for leather goods and sampled mint tea. William was presented with a complimentary leather belt suggesting that perhaps we had not been as good at the bargaining as we might have been. I was concerned when Elizabeth was approached by a young steward on the ship offering drugs, but having had lessons that term on the dangers thereof she knew what was going on and walked away, smart enough not to get involved. We all loved the basket trip down from the heights in Madeira, while in Gibraltar, as well as the apes on the Rock, we encountered the England football team; the children obtained Gordon Banks's autograph, which they took triumphantly back to school the next term. It was a delightful holiday.

I joined Stowmarket Hockey Club soon after leaving school in 1949 and usually played half-back. In a small club adaptability is key, though, and I have played in every position, even in goal. We had no pitch of our own and so all our games were away. In the 1950s we played several Service teams including RAF Stradishall, HMS *Ganges* and the Suffolk Regiment based at Bury barracks, but as the armed forces closed these bases we had to find other opposition. Most towns had hockey clubs fielding one or two teams, as did many of the local firms, such as Cranes, Fisons and Ransomes. We played them all.

Our team was made up mostly of local farmers with a sprinkling of agricultural engineers and lawyers, and a grocer, all of whom were or became good friends. Among them was Keith Bullock, who had been at school with me and farmed in Cotton, and John and Rowley Miles, farmers from Great Ashfield, who were backs. The pitches then were frequently quite rough and the ball was more often in the air than on the grass, but the Miles brothers were adept at hitting it wherever it was. Dick Forrest, another farmer, played centre-forward. The three Nunn brothers, Edward,

Bill and Peter, were also in the team and with their father made up Frank J. Nunn and Sons of Elmswell, Massey Ferguson dealers with whom we did a good deal of business. The Nunn brothers had been a few years ahead of me at Culford School and before my first match for Stowmarket Bill took me to one side and warned me that play may be a bit tougher and rougher than I was used to. Another member of our team was Manning Prentice, a Stowmarket solicitor. He was maybe slightly eccentric but was also a highly regarded amateur astronomer. One night he spotted a hitherto unknown star in the constellation of Draco, which he was able point out to the Royal Astronomical Society. He also spent a great deal of time monitoring meteors, work for which he was internationally recognised. Manning drove an open-topped tourer with the words 'Dieu et mon droit' inscribed on the sides and on one occasion played hockey in his bare feet, I assume because he had left his boots behind.

I continued to play hockey on Saturday afternoons in the autumn and winter after I married, and Mary and the children would occasionally watch from the touchline. In those days we did not have the mid-week training or fitness sessions that clubs seem to insist on nowadays. I was lucky to be playing at a time when it was more important to enjoy the game than to worry about winning. I do not think I could have fitted in the hockey around farming and family life if we had all taken it too seriously but there was much camaraderie and friendship which developed, not only among our own team members but also with players from opposing teams whom we would meet year after year. One of the opposing teams was in the habit of offering us a half bottle of Grand Marnier to share round at half time, which was always gratefully received in the cold of winter but probably did nothing to improve our game.

Easter was a time for hockey festivals and in 1964 Mary, the children and I stayed at Clacton for the weekend hockey festival. When I turned up on the pitch for the first match on the Sunday morning, my teammates asked, 'What were you up to last night?' I was at a loss to know what they meant. It had been the night when the Mods and Rockers had a set-to on the beach and almost set Clacton alight, making national headlines. We had slept through it all and were peacefully unaware of any fracas.

One Saturday afternoon in 1967, with Elizabeth and Christine in the car, a Ford Zephyr, I picked up John Gooderham and Gilbert Bradfield to go to a hockey match in Ipswich. I was overtaking a car on the A140 near the Coddenham turn off when suddenly the driver turned right in front of me. Swerving to avoid him I careered off the road through a fence, the car falling 4ft over a drop before landing close to a fair-sized yew tree. It was the days before seat belts but fortunately little damage was done to the occupants, those in the back being jammed tightly together. Clambering out and slightly stunned we wondered what to do next. As luck would have it, shortly afterwards some other team members drove by so John and Gilbert were able to continue the journey with them and play the match, while I and the girls hitched a lift back to the Rosery at Little Stonham where Mary's brother William and his wife, also named Mary, then lived. (When William and Mary married in 1969, my in-laws moved out of the Rosery into a bungalow they had built on their farm.) After a cup of tea for me and Smarties for the girls, the shock began to wear off and we started to feel better. My parents-in-law collected Elizabeth and Christine and drove them back to Harpsfield, while brother-in-law William drove me to the doctor for a check over. Mary was surprised and worried to see the girls home without me. 'Where's David?' she asked. All was explained and I was back a little later with just a few stitches in my lower lip. We had been lucky but the car was a write-off.

As time went on, we had difficulty in finding enough players for Stowmarket to keep going and the club folded in the 1970s. I was invited to play for the Norwich Exiles which I did for the next 20 years. There were occasions when I would be playing hockey at Eaton Park in Norwich while one or more of our daughters was playing lacrosse for their school, Runton Hill, against Norwich Ladies on another pitch in the park. Once or twice I played in Lancashire for my son-in-law's team. It had a similar, friendly atmosphere and the banter there was much the same as in Suffolk, even though the dialect made some of it incomprehensible.

I never spent much time just watching the game although I did enjoy seeing grandson Andrew playing school hockey on AstroTurf at Clifton College in Bristol, and again when he was at Cambridge. Many other members of our extended family play hockey and I have happy memories of mixed matches on Boxing Day when the Blacks played 'the rest of the world' (those friends we could manage to rustle up) on the local football pitch, sometimes with as many as 16-a-side, with hot drinks and much bonhomie afterwards. We now have great-grandchildren taking up the hockey stick and it remains to be seen whether they get as much pleasure from the game as I did, though it is a different sport nowadays. The rather rough-and-ready pitches have given way to AstroTurf and its successors. Hockey is now much more intense, and the league system has given it a far more competitive edge.

As a family we have enjoyed many skiing holidays. I first skied with my brother Robert in 1950 in Grindelwald, on a very basic trip. We travelled in a freezing cold railway compartment overnight and skied as best we could in corduroy trousers. The room in the chalet we rented was heated with a 'Tortoise' stove which did not quite stop the frost from forming inside the windows overnight. Quite by chance while there we met Basil Ambrose and David Gooding, both Old Culfordians who were three or four years older than us but barely more competent. The cold and simplicity of the trip did not put us off completely because early in 1957, before our weddings later that year, Robert and I took Mary and Miriam to Kandersteg.

I did not ski again until 1973 when Mary and I felt able to take the four children to a small resort, Fulpmes, in Austria.

It was there in the Hotel Lutz that we first met the Grant family. John farmed just outside Boston in Lincolnshire and he and his wife Margaret had four children of a similar age to ours; we got along famously. We had many similar interests, including playing bridge, and I am pleased to say that we forged a friendship that has lasted to the present day in both generations.

I was an enthusiastic skier but never as proficient as I would have liked, despite all the lessons I have taken. A week a year was not enough, but I always found it a challenging and pleasurable experience. The children too enjoyed skiing and the mountain air, and it was great to see them able to excel and do better than their parents. Elizabeth was the first to give up our annual visits to the Alps. It is ironic that she and Willie, her husband, now own two apartments and spend five months each year living in the ski resort of La Tania in the Courchevel valley. They enjoy the skiing, manage several other apartments, including one that Mary and I bought in 2008, and attend to the needs of many of the extended family and friends who find their way there in the season. Mary became less enthusiastic about downhill and took to cross-country skiing. I occasionally accompanied her, but I have always enjoyed the thrill of the wind whistling past my ears and experiencing the feeling of being high up among those majestic snow-covered Alps. Each time I have returned home uplifted and feeling fitter than before the holiday.

All too quickly, or so it seemed to me, our children grew up, finished their schooling, travelled, studied. The girls forged careers elsewhere while William joined the farming business. They all married and had children. Their exploits could fill another book, but I am not the one to write it. Each summer they would descend for 'Suffolk Week' with their young families coming from Lancashire, London and Bristol, the next generation of cousins having the chance to spend time together and to experience some of the fun our children used to have in the garden and on the farm. It was super to see the grandchildren take to the ski slopes as we enjoyed several multi-generational chalet holidays together in the early 1990s, and now it seems that the great-grandchildren have also caught the ski bug.

Overall, we have been very fortunate as a family, but we have also had to face two tragedies, the first when my brother Thomas died in 1989, and then when our grandson, Edward, Elizabeth and Willie's son, had a terrible accident at the age of 17 while in Interlaken. He was airlifted to hospital but could not be saved. His parents showed enormous courage as they faced their loss and donated his organs for transplant to help others. A large family provides a great comfort in times of sorrow and sadness. Friends, too, play their part and we have been supported and encouraged by so many kindnesses over the years.

We feel privileged that, 30 or so years after leaving home to lead their various lives, the girls returned to live in Suffolk with their husbands, and we may now spend more time with them and their families. We have eight grandchildren and, so far, five great-grandchildren. Mary and I have enjoyed creating the garden over the years and it gives us enormous pleasure to share it with family and friends. Together with the pool and tennis court, it still provides a hub for a get together and great entertainment for lively youngsters.

The family has always been involved in village life and we have been happy to play our part alongside many other enthusiastic and willing volunteers. When I was a child, we hosted village parties on the farm for special occasions, such as the King's Jubilee in 1935 and the Coronation of King George VI in 1937. Father was active on both the parish and district councils for many years. The period just after World War II was a particularly busy time for making decisions, as council houses were being built, water was being piped to the villages, electricity was being supplied and sewage schemes were being introduced to replace the old honey carts. Mother was president of the village Over-Sixties Club which met weekly in the village hall. She would be there on Monday afternoons, ensuring that Scrabble was included on the schedule alongside whist and beetle drives.

We were able to provide the land for a football pitch in Bacton during the 1960s and for the scout hut beside it, and then later in 1976 some land for a bowls club. Father was invited to be president and rolled the first wood on the green. The connection continues to this day, and Mary and I are still invited to the annual bowls club dinner where she and Rosemary Jeffries present the prizes. Rosemary and her husband Philip run a car dealership in the village and are generous village benefactors, sponsoring the bowls club, the flower show and the football club. Another notable villager, Ernie Pearce, part of a long-established Bacton family, was a leading light in the bowls club and was one of those willing volunteers involved with the club from its inception. He ran his own painting and decorating business and for many years did all the painting and decorating we needed both at Harpsfield and around the farm. A prominent member of the local Lions club, Ernie raised a great deal of money for charity. He was a keen Ipswich Town Football Club supporter and, on several occasions, took William to see the team play at Portman Road.

Mary, too, has done her bit in the village and elsewhere. Over the years she has supplied a constant stream of flower arrangements to the church, polished pews, baked cakes and made sandwiches for the flower show and numerous other goings on. She remains a member of the Women's Institute and of the church House Group. The 1st Bacton and Cotton Brownies was formed in the 1960s, with Mary as Tawny Owl and Mary Dodds as Brown Owl. Our girls, together with Mary Dodd's daughter Louise, were keen members. Mary was also much involved in the Stowmarket branch of Action Medical Research, as secretary to chair Phoebe Williams of Haughley Park, before becoming chair herself. Fundraising highlights of her term include hosting Trevor MacDonald of *News at Ten*, as the guest speaker at a luncheon, and persuading the King's Singers to perform at the Corn Exchange in Ipswich.

I was asked to become treasurer to St Mary's Parochial Church Council (PCC) in the 1950s. The churchwardens at the time were two brothers, Tom and Creasey Gooderham, but not long after I had joined the PCC Tom was killed in a tragic accident with a combine harvester on his own farm. I was appointed churchwarden in his place, a post which I held for the next 50 years and through which I was involved in a variety of different projects. In that time there were nine different rectors, various reorganisations of the parish and the formation of new benefices, the current one having six parishes.

When I started as churchwarden, Frank Evans was rector, having succeeded the Revd Bowden in 1948, and he and Creasey initiated me into the position. When Creasey died in 1974 I was joined by Bill Stannard, from an old Bacton family, who had been away teaching but had retired to the village with his wife Brownie. They both had an enthusiastic love of Bacton and, as I was actively engaged in farming, he did much of the churchwardenly duties relating to the preparation for services while I took responsibility for overseeing repairs and maintenance.

I remember one Sunday morning, with the congregation all assembled for 8 o'clock communion, there was no sign of the rector, a popular and much-loved incumbent. After some time, it was suggested that I should go to find him. I went across the road to the Rectory but could not make him hear. As the door was unlocked, I went inside, but there was still no sign of him. Slightly concerned, I went upstairs to his bedroom where I found him sound asleep. I had not the heart to wake him as he looked so peaceful. I returned to the church, told the congregation, said a prayer and we all departed. A somewhat similar situation arose with his successor but this time I was encouraged by the congregation to lead those parts of the service which a lay member could. I had no glasses with me so thought that was that. However, one member of the congregation insisted I wear her spectacles so that I could proceed. I am told that Dame Edna Everage conducted a passable act of worship that morning.

The repair and maintenance of the church building has been a constant struggle. Mary said she was frequently concerned when thinking of me, up a ladder, sorting out some problem or other at the church or overseeing work being done there. The lead covering to the side aisles was replaced with roofing felt which, after 40 years, was in its turn replaced with lead again. Fashions and advice change. In 1972, with James Mair, an enthusiastic rector, we raised sufficient funds to have the bells recast and a new one added, bringing the peel to six bells. In my time, other projects have included the restoration of eroded stonework and the refurbishment of the Doom painting above the chancel arch. The chancel roof itself was repainted in memory of my brother Thomas and the Revd Robin Jack, in 1995. I was up and down the scaffolding on many occasions to see how the work was progressing and to supply buckets of water and whatever else was needed. I retired as churchwarden in 2017 but continue to help where I can.

Another role of the churchwarden was to chair the Bacton Charity Trustees. The committee had to decide who in the village should receive support from the small amount of funds available, but it often concerned me that despite our best efforts we might not necessarily know who was most in need. More recently, as the public has become much more concerned about local heritage, several parishioners, not all regular churchgoers, came together to form the Friends of St Mary's Church, Bacton. They have helped enormously with organising fundraising events such as concerts, metal detecting days, and an annual church lunch which is held the day after the flower show in early August. Following the lunch there is an auction of promises, another useful money raiser, and for many years I would encourage whoever I could, local businesses, villagers, family and friends, to offer items or 'promises' for auction, and persuade a local auctioneer to lead the sale. The flower show was moved from our own Show Meadow to the Bacton Middle School in 1961 soon after we built Harpsfield, but now it is held in the meadow opposite the Manor in Bacton. These events, together with contributions from charitable organisations, have enabled us not only to carry out maintenance work on the church and to pay the quota to the diocese, but also to install a WC and a community room in the back of the church, completed in 2011. This has made a more convenient and warmer place to hold meetings and for serving refreshments after services and events. Simultaneously, the organ was refurbished, extended and moved above this new room to an organ loft, with funds largely and generously provided by our long-time organist and local doctor, Peter Franks.

Harpsfield, 2021

Mary and I at a Buckingham Palace Garden Party, July 1991

Mary and I in Alaska, 2016

St. Mary's Church, Bacton, Suffolk

Doom painting, St. Mary's church, Bacton, restored in 1968

CHAPTER 10

REFLECTIONS

As I look back at my working life, the way that my brothers and I were inducted into the family farming business seems to have been rather a casual way of deciding our futures. It worked well enough then because matters were decided around the Somerville kitchen table. Decisions regarding the farm continued in this way for some years even after both Robert and I were married. We both recognise how lucky we were to have been offered partnerships in the business early on, me in 1951 just two years after I had left school, and Robert a couple of years later. At that stage, Grandad owned two-thirds of the partnership and Father one-third. By 1956 Grandad was in a slow decline and we were advised by our accountant, Bob Taylor, that there would be benefits if the partnership were to become a limited company with shares. It was agreed, and as a result of this change Grandad passed over assets and finished up holding 5,000 shares, Father held 15,000 shares and each of us boys 10,000. Thomas's shares, for he was then aged only 10, were held in a Trust. We were very much a family business then as now. The Articles of Association adopted in 1956 have stood the test of time with only the odd tweak since to comply with changing legislation.

As time went on and another generation appeared it seemed reasonable that the boys[1] should have the chance to join the business. James, Robert's eldest son, took a degree in

---

[1] It seems strange but I'm afraid that is what we thought in the 1970s, although at different times the girls had holiday jobs on the farm and Alison also worked with the pigs for a few months between school and University.

agriculture at the University of Reading and then spent time working on other farms including a spell with John Hitchcock at Ringshall. In 1983 I was pleased to be able to ask him to come to work with us, particularly to deal with the financial side of the business and to oversee computerisation. His brothers, Andrew, David and Peter decided to seek their futures elsewhere. My son, William, some five years younger than James, worked on other farms in the UK and in Australia and took an agricultural course at Writtle before he too joined us in the business. His area of responsibility was the maintenance of the existing properties and the development of new ones as we expanded the pig business and developed the AI operation.

Change has been with me all my life and the multitude of crops that we grew when I left school has steadily diminished. This was either because they became unprofitable, because our heavy land was not suitable, because we did not grow enough of the crop to benefit from scale, or because the demand dried up. The fruit trees were the first to go, then the bulbs and the cut flowers; on the arable side, out went the potatoes and vegetables, then hay, the clovers and grass seeds, with cocksfoot staying on longer until black grass eventually put a stop to that. We increased the acreage of wheat as well as the sugar beet, and introduced oilseed rape. The onions and the rhubarb continued until 1993 when, after careful consideration, we decided to stop growing these crops which meant that, regrettably, we had to make the gang of ladies redundant. Robert, who had overseen the horticultural side of our business, eased into retirement, which enabled him to spend more time at Chadacre and Otley, the two agricultural educational facilities in Suffolk, where he was a governor and chairman for over 20 years.

In 1998 we celebrated 100 years since Grandfather had come from Scotland to Red House Farm. We held a formal dinner on the Saturday night and one of the sheds was turned into a fun fair open to all the staff and their families the next day. What with bouncy castles, slides, bucking bronco and numerous other activities, as well as a hog roast, a great weekend was had by all, and Mother and Father were there to enjoy the company and reminisce. That same year, Mark, Tom's elder son, joined us on the farm. After completing a degree in agriculture at the University of Newcastle, Mark had spent a couple of years working for Peninsular Pigs at their headquarters in Frome, Somerset and then completed a six month 'stage' in the agricultural department of the European Union in Brussels.

On the morning of 4 June 1999 Father rang, asking me to see him at The Grange, the nursing home where he had been living for just over a year. He was wanting to know what was happening on the farm and what I had seen at the Suffolk Show on the previous two days. We chatted about this and that, and then he said he would like to go back to bed. The staff at the nursing home felt he was none too good and called Mother and the rest of the family. A little later, with the family around him, he passed away peacefully, just short of his 99th birthday.

In 2001 foot and mouth disease was running rife throughout the country and concern about its spread caused the Suffolk Show and other county shows to be cancelled. Fortunately we avoided the disease on the farm, but it cost us dear, as no animal movements were permitted off the farm and pig prices nationally were in a parlous state. Our pig operation was continuing to go through difficult times and pig prices that year had dropped by 40 per cent between March and September. James, with Simon Guise, our pig manager, was desperately trying to reduce costs and improve efficiencies. There is nothing like hard times to focus minds. Loss-making units were closed and there were inevitable redundancies.

Having by that time reduced the number of different crops we were growing on the farm, we all agreed we could do with some other business to balance our dependence on the pigs. I was by then spending less time on the day-to-day running of the farm and decided to visit and speak to anybody who I thought might have a bright idea as to what this additional business might be. Nothing quite seemed to fit the bill until I met Brian Belcher. Brian was into house building and repairing and, seeing an opportunity, had bought part of the old Stanton aerodrome. He let out some of the buildings, refurbished others and set up several businesses. Among these was a cabin and container-hire operation which his son, Michael, had only recently sold to a Yorkshire-based company. However, he knew of a similar business, run by Bill Nay from Offton, which had been renting out cabins and small tools. I was led to believe that Bill could be persuaded to sell the cabin-hire operation so that he could expand on the tool-hire side. It looked promising: we had the space to store cabins, it was a business that could develop and one that I hoped could make good use of Mark's talents. Through my membership of the 1975 Club I knew that one of our members, Simon Scott, and his family, were also in the cabin business and Mark and I paid a visit to Chester to see them. There we met with Simon's brother Justin and manager Glynn Ball who were very open about the finances of their business. They suggested that, although there were already some national cabin-hire businesses covering the country, there was room in the market for non-national companies, particularly if they could make a quick response and would be prepared to take on more complicated and unusual jobs. The upshot was that we bought the business, renaming it Portable Space and made it Mark's responsibility.

Robert and I were both by now approaching 70. James was managing director, William, director in charge of the service side of the business and Mark, commercial director. Slowly but steadily it was becoming obvious that the informal way in which I had been happy to run the business was no longer fit for purpose. We had many discussions among the family as to how to proceed for by then all 11 members of the fourth generation (my children's generation) were shareholders irrespective of whether they worked in the business or not. Dividends were an effective way of providing them each with an allowance throughout their student days.

In retrospect I wonder how we all managed to work together so well for so long. All family relationships experience stresses and strains to a greater or lesser degree and our family has been no different. I remember Father telling me that at one time in the late 1920s he was seriously considering leaving the farm and emigrating; the future for UK farming looked bleak and the relationship between him and his father was strained. In the event he decided to stay, which is why I am here and able to write this piece. Perhaps his frustration, all those years before, influenced the way he handed on responsibility to me, Robert, and then Thomas early on in our working lives on the farm. The great thing is to be able to face any challenges and hopefully rise above them to emerge stronger and more resilient. My mother was a great one for quoting St John: *'Love one another as I have loved you'*[2] and, perhaps, she could be described as the family 'glue'. Each weekday morning, she baked a round or two of scones and was never happier than when we dropped in for coffee at Somerville. This ensured regular conversations around the kitchen table, even if sometimes we were running from one job to another. Mother also felt it important to treat each member of the family equally, convinced that this was a recipe for family harmony, and I have tried to let this view guide me. As we progressed into the second century of the business, we needed a solution for the next generation.

In 2008, through contacts that James and Mark had made at Cranfield, we learnt of Peter Leach, who specialised in family businesses and had wide experience of advising families passing through similar changing times. He was a great help, reassuring us that we were not unusual. In hindsight we should have called him in sooner. He was a good communicator, pointing out strengths and weaknesses, and interviewed all family shareholders individually, thus gaining an insight into what they really thought rather than

---

[2] John 13, 34.

what they were prepared to say in open discussions. His intervention led to us making a formal separation between the interests of the business and those of the family, with a written constitution, and with the formation of a Family Council to provide the formal link between the two. An executive board was appointed to oversee the running of the business and to set budgets, and the executives were to have formal contracts. I became chairman of the board and of the Family Council but took a step back and was no longer involved with the day-to-day running of the business.

We had already been considering appointing someone from outside the family to join the board and Peter Leach helped accelerate that process. Gordon McIntyre, a Scot and a professional agricultural advisor living locally, was much recommended, and although he chose not to become a non-executive director he became an advisor to the board. He does not pull his punches and has given us quite a shake-up. The family has much to thank him for over the nine or 10 years that he has been advising us. We are a much leaner and more focused farming operation than before and each department now has its own budget. Both William and Mark felt some frustration in their roles and William left the business in 2011. The cabin-hire business, Portable Space, grew well under Mark's management but he was keen to run his own business and, after lengthy deliberations with the Family Council, bought it out in 2012 to operate as a separate enterprise outside the rest of the family business.

The wider family have been encouraged to take an interest and learn more of the business. The day of the AGM gives a chance for a get together, to learn more of the business and farming activities, and, after the formalities, for a catch up between the ever-expanding branches and generations of the family, finishing with a dinner and a knees-up or home-grown entertainment in the evening.

What happens next? In 2016 the fifth generation were each given 100 shares in the company by their respective grandparents, and David Black & Son Ltd remains an entirely family-owned business. I have continued to reduce my commitment, resigning as chairman in 2019, but remain on the board where I hope I am still able to make a useful contribution. I suggested that my nephew, Ian David, one of Robert's sons, should become chairman and we appointed both the first member of generation five and the first non-family member to the board. The former is my grandson, Andrew Bailey, who, outside our business, is working in management consultancy, and the first non-family member is our finance director, Jay Hawes.

Since I left school there have been great changes to our way of farming and the crops we have grown. One may make a decision today, but it may be many years before the effect of that decision becomes obvious. Seventy or 80 years ago, when the country needed us to increase production and efficiency, we cut down the hedges, filled in the ditches, made more machinery-friendly fields. Now that we have upped production the pressure is on to return some of the land to its former state. We are already planting more trees and hedges and with the help of a volunteer conservationist we have seen an increase in the number and variety of wild birds on the farm, including the now-rare grey partridge. Food production appears to be taking second place to 'public good' which has yet to be clearly defined. I have lived through interesting times and there are interesting times ahead. Since I left school there have been great changes to our way of farming and the crops we have grown. We have tried to keep up with the changes, make a living for the family, play our part in the locality and produce food for the nation. What we produce and how we produce it is governed mostly by consumers and what they are prepared to pay. The supermarkets, government regulations and the weather also come into the mix and whatever we do we have to do it more economically and with a smaller carbon footprint. The business will have to continue to adapt and alter as we face the future.

I have only been able to mention a few staff by name, but the business would not have thrived without the support, dedication and work of everyone involved. Without them, nothing I have recounted would have been possible and I thank them all. There have been times when several fathers and sons were working on the farm simultaneously, and many of our loyal staff have received long-service awards in recognition of their commitment.

One wonders what the business might look like in 100 years' time. It has provided me with a long and fulfilling life and has been as much my hobby as my work. I would like to think it could survive and prosper over the next 100 years and provide a springboard for opportunities and an inspiration for more generations of the Black family.

Poplar trees planted near stream at Wickham Skeith in the 1960s. The expected market for matches had disappeared by the time these trees had grown large enough, but they enhance the landscape

Brooding clouds over stubble field at Bacton

How times change. By 2014, this Vervaet Beet
Eater 617 harvester could do the work of 50 men

Involving the next generation. Family Assembly, 2015,
Farming & Wildlife theme. Members of
the 5th generation examining owl pellets

Sunset over the slurry lagoons

Lightning Source UK Ltd.
Milton Keynes UK
UKHW010650150422
401605UK00001B/5